AF404723

A Practical Introduction to Beam Physics and Particle Accelerators
(Third Edition)

Online at: https://doi.org/10.1088/978-0-7503-4039-7

A Practical Introduction to Beam Physics and Particle Accelerators
(Third Edition)

Santiago Bernal

Institute for Research in Electronics and Applied Physics, University of Maryland, College Park, MD, USA

IOP Publishing, Bristol, UK

ISBN   978-0-7503-4039-7 (ebook)
ISBN   978-0-7503-4037-3 (print)
ISBN   978-0-7503-4040-3 (myPrint)
ISBN   978-0-7503-4038-0 (mobi)

DOI   10.1088/978-0-7503-4039-7

Version: 20221101

IOP ebooks

British Library Cataloguing-in-Publication Data: A catalogue record for this book is available from the British Library.

Published by IOP Publishing, wholly owned by The Institute of Physics, London

IOP Publishing, No.2 The Distillery, Glassfields, Avon Street, Bristol, BS2 0GR, UK

US Office: IOP Publishing, Inc., 190 North Independence Mall West, Suite 601, Philadelphia, PA 19106, USA

In memory of Martin Reiser (1931–2011)

Contents

Preface to the third edition   x

Acknowledgements   xi

Author biography   xii

1   Rays, matrices, and transfer maps   1-1

1.1   Paraxial approximation   1-1

1.2   Thin lenses   1-3

1.3   Thick lenses   1-4

1.4   Transfer maps   1-6

1.5   Computer resources   1-8

     References   1-9

2   Linear magnetic lenses and deflectors   2-1

2.1   Magnetic rigidity, momentum, and cyclotron frequency   2-1

2.2   Solenoid focusing   2-2

2.3   Quadrupole focusing   2-5

2.4   The Kerst–Serber equations and weak focusing   2-8

2.5   Dipoles and edge focusing   2-10

2.6   Computer resources   2-11

     References   2-13

3   Periodic lattices and functions   3-1

3.1   Solenoid lattice   3-2

3.2   FODO lattice   3-3

3.3   Lattice and beam functions   3-7

3.4   Uniform-focusing ('smooth') approximation   3-11

3.5   Linear dispersion   3-12

3.6   Momentum compaction, transition gamma, and chromaticity   3-13

3.7   Computer resources   3-15

     References   3-15

4   Emittance and space charge   4-1

4.1   Liouville's theorem and emittance   4-2

4.2   The Kapchinskij–Vladimirskij (K–V) and thermal distributions   4-5

4.3 Thermodynamics of charged-particle beams?   4-7

4.4 The K–V envelope equations and space-charge (SC) intensity parameters   4-9

4.5 Incoherent space-charge (SC) betatron tune shift   4-12

4.6 Coherent tune shift and Laslett coefficients   4-16

4.7 Computer resources   4-18

References   4-18

5   Beam (sigma) matrix and coupled optics   **5-1**

5.1 Solenoid focusing revisited   5-1

5.2 Skew quadrupole   5-5

5.3 Beam (sigma) matrix   5-6

5.4 Coupled optics   5-8

5.5 Angular momentum and the envelope equation in solenoid   5-10

5.6 Round-to-flat (RTF) and flat-to-round (FTR) beam adapters   5-11

5.7 Computer resources   5-15

References   5-16

6   Longitudinal beam dynamics and radiation   **6-1**

6.1 Radio-frequency (RF) linacs   6-1

6.2 Beam bunch stability and RF buckets   6-4

6.3 Synchrotron radiation   6-6

6.4 Insertion devices and free-electron lasers (FELs)   6-9

6.5 Longitudinal beam emittance and space charge   6-11

6.6 Computer resources   6-18

References and additional reading   6-19

7   Envelope matching, resonances, and dispersion   **7-1**

7.1 Cell envelope FODO matching   7-1

7.2 Source-to-cell envelope matching   7-7

7.3 Betatron resonances   7-8

7.4 Betatron resonances and space charge   7-12

7.5 Dispersion and space charge   7-14

7.6 Computer resources   7-17

References   7-19

8 Linacs and rings (examples), closed orbit, and beam cooling — **8-1**

8.1 Examples of linacs — 8-1

8.2 Examples of rings — 8-3

8.3 Closed orbit and correction — 8-7

8.4 Beam cooling — 8-9

8.5 Computer resources — 8-11

References — 8-11

9 Small machines and scaled experiments — **9-1**

9.1 The University of Maryland Electron Ring (UMER): a storage ring for space-charge research — 9-2

  9.1.1 UMER lattice and injection — 9-2

  9.1.2 Closed orbit — 9-4

  9.1.3 Betatron tunes and space charge — 9-4

  9.1.4 Betatron resonances and space charge — 9-5

  9.1.5 Soliton trains and multistream instability — 9-6

  9.1.6 Nonlinear optics — 9-7

9.2 Small Isochronous Ring (SIR): space-charge effects in the isochronous regime — 9-8

9.3 Integrable optics and other physics in IOTA — 9-11

  9.3.1 IOTA lattice and diagnostics — 9-11

  9.3.2 Nonlinear integrable and quasi-integrable optics — 9-13

  9.3.3 Optical stochastic cooling — 9-15

  9.3.4 Space-charge compensation and other experiments — 9-15

9.4 Fixed-field alternating-gradient accelerators (FFAGs): lessons from EMMA — 9-16

9.5 Beam stability and betatron resonances: Paul traps as model accelerators — 9-22

9.6 Computer resources — 9-29

References — 9-30

Appendix A: Computer resources and their use — **A-1**

Appendix B: Accelerator magnets — **B-1**

Preface to the third edition

The third edition of this book maintains the same general character as the first two editions, which is a brief exposition with emphasis on numerical examples. However, we have expanded the discussion with new theoretical topics as well as applications. We have added two chapters (5 and 9) and an appendix B. The new chapter 5 is devoted to the beam (sigma) matrix and circular optics, major areas of accelerator theory and research. Chapter 9 is an extension of chapter 8 (which covers examples of linacs and rings, closed orbit, and beam cooling) but with examples of smaller machines, including Paul trap alternating-gradient lattice simulators. The chapter also introduces a few new concepts, such as instabilities in the isochronous regime, nonlinear integrable optics, and fixed-field alternating-gradient accelerators (FFAGs). The new appendix B, on the other hand, is a summary of accelerator magnet theory: multipole expansion, linear and nonlinear magnets, and the effective hard-edge model of focusing magnets.

Furthermore, we have revised and updated appendix A, and learned that the lifetime of working hyperlinks is shorter than we had thought. Finally, we have qualified or made corrections to a few theory statements or equations related to the Courant–Snyder theory, dispersion, and beam cooling, among others.

The examples and computer exercises comprise basic lenses and deflectors, fringe fields, lattice and beam functions, beam (sigma) matrix, circular optics, synchrotron radiation, beam envelope matching, betatron resonances, dispersion, transverse and longitudinal space charge, closed orbit, and beam cooling. Appendix A gives the reader a brief description of the computer tools employed and concise instructions for their installation and use in the most popular computer platforms (Windows, Macintosh, and Ubuntu Linux). Hyperlinks to websites containing all the relevant files are also included. An essential component of the book is its website https:// ireap.umd.edu/clark/faculty/1273/Santiago-Bernal (which is actually part of the author's website at the University of Maryland). It contains files that reproduce the results given in the text as well as additional material, such as technical notes and movies. We will add new or updated material as it is developed. Although we have chosen Mathcad for most examples, we will add Matlab and Python scripts in the near future.

Acknowledgements

The discussions we have had over the years with members of the University of Maryland Electron Ring (UMER) group and the study of Prof. Martin Reiser's book *Theory and Design of Charged-Particle Beams* were the inspiration that led to the present book. Special thanks to our colleagues and collaborators at the University of Maryland: Max Cornacchia, Brian Beaudoin, Irv Haber, Rami Kishek, David Sutter, Dorothea Brosius, and Professors Tom Antonsen and Pat O'Shea. We also acknowledge the generosity of Chris Allen (Oak Ridge Natl. Lab), Professor Alex Dragt (University of Maryland), Rob Ryne (Lawrence Berkeley Natl. Lab.), Hui Li (Confer Technologies, Inc.), and Valter Kiisk (University of Tartu, Estonia) for granting us permission to include their computer codes in our website and/or for assistance in their installation or use. Special thanks to Laurent Deniau (CERN) for pointing out errors in our example of the MAD-X code in the appendix of the second edition, and for providing us with a better version.

We mention again several former students who were instrumental in preparing resources for the second edition: Heidi Baumgartner Komkov, Levon Dovlatyan, Samuel Ehrenstein, William Hartmann (R.I.P.), David Matthew, Sarvenaz Memarzadeh, and Kiersten Ruisard. For the third edition, we thank Levon Dovlatyan (Raytheon Technologies) and Eduard Pozdeyev (Fermilab) for granting us permission to use material from their PhD theses. In addition, thanks to Giulio Stancari (Fermilab) for directing us to publications on IOTA, Traci Marc (Fermilab) for the license to use material from Fermilab's website, and Fiza Mulla (Freedom High School) whose summer project in 2020 (via Zoom) on scaling accelerators was the inspiration for the new chapter 9.

We also acknowledge the patience and support of our family and the publishers. Last but not least, we acknowledge the continuing financial support of the Offices of Science and High Energy Physics of the U.S. Department of Energy.

Author biography

Santiago Bernal

Santiago Bernal obtained a BS in physics from the Universidad Nacional de Colombia in Bogotá, Colombia in 1981, an MS in physics from Georgia Tech in 1983, and a PhD in physics from the University of Maryland, College Park in 1999. He worked for his PhD under the direction of the late Prof. Martin Reiser. Before his graduate studies at Maryland, Bernal taught college physics and math in Colombia and Puerto Rico. Bernal joined the UMER group in 2000 as a postdoc, later becoming a research scientist at the Institute for Research in Electronics and Applied Physics (IREAP). He was involved in the design and construction of UMER and has been a leading experimentalist in the project. Bernal helped organize and conduct an experimental course in UMER for the 2008 US Particle Accelerator School. He is the junior coauthor with Charles L Joseph (Rutgers University) of *Modern Devices: The Simple Physics of Sophisticated Technology* (Wiley, 2016). Chapter 22 of *Modern Devices* formed the basis of this book. In addition to beam and accelerator physics, Bernal is interested in statistical mechanics and the educational aspects of physics.

IOP Publishing

A Practical Introduction to Beam Physics and Particle Accelerators (Third Edition)

Santiago Bernal

Chapter 1

Rays, matrices, and transfer maps

To understand particle accelerators and their components, we need to study and solve the equations of motion of charged particles in external electromagnetic fields. In their most general form, these equations contain terms involving all orders of the coordinates and velocities of the particles [1, 2]. However, if the particles are confined to small distances from a *reference orbit* and have small angles relative to the same orbit, a *linear equation of motion* resembling a simple pendulum (or mass-on-a-spring) equation results. This is the basis for the *paraxial approximation* presented in section 1.1. The 'spring constant' in the equation of motion can be generalized to include a piecewise *focusing function* whose form depends on the types of external fields present. In addition to the paraxial approximation, the focusing elements of particle accelerators can be treated as *thin* or *thick lenses*, depending on whether the focal length is large or comparable to the physical axial extent of the external fields. We discuss thin and thick lenses in sections 1.2 and 1.3. For both thin and thick lenses, a *matrix* description of particle motion is both convenient and powerful. Finally, in section 1.4 we mention the existence of powerful methods based on Lie operators that address higher-order particle tracking.

1.1 Paraxial approximation

The linearized equation of motion for a charged particle in the presence of external fields represented by the focusing function $\kappa(s)$ is

$$x''(s) + \kappa(s)x(s) = 0, \tag{1.1}$$

where primes indicate derivatives with respect to the axial distance 's' measured along the reference trajectory. For now, we consider just one of the transverse coordinates for particle motion, the horizontal component (or the radial component in a local curvilinear coordinate system).

Note that equation (1.1), known as Hill's equation, is a homogeneous, second-order ordinary differential equation (ODE) with a coefficient, the focusing function, which is not generally constant. The solution depends on the initial conditions $x(0)$, $x'(0)$, and on the form of $\kappa(s)$, but not on any higher orders of either $x(0)$ or $x'(0)$. This latter condition is the essence of the *paraxial approximation*; it is the equivalent of using θ instead of $\sin\theta$ in Snell's law of standard optics.

Here and in subsequent chapters we will discuss the solutions of Hill's equation and some of its non-homogeneous versions in some detail; additional considerations can be found in many textbooks (e.g., [2]). The general solution of equation (1.1) can be written as a superposition of *cosine-like* $C(s)$ and *sine-like* $S(s)$ functions

$$x(s) = x(0)C(s) + x'(0)S(s),$$
$$x'(s) = x(0)C'(s) + x'(0)S'(s), \tag{1.2}$$

such that $C(0) = 1$, $S(0) = 0$, $C'(0) = 0$, $S'(0) = 1$. In matrix form, equation (1.2) reads

$$\begin{bmatrix} x(s) \\ x'(s) \end{bmatrix} = \begin{bmatrix} C(s) & S(s) \\ C'(s) & S'(s) \end{bmatrix} \begin{bmatrix} x(0) \\ x'(0) \end{bmatrix}. \tag{1.3}$$

In a more compact form we can write

$$[x(s), x'(s)]^T = R[x(0), x'(0)]^T. \tag{1.4}$$

The space spanned by the coordinates $x(s)$, $x'(s)$ for an ensemble of particles is called *trace space* or, more commonly, *phase space*. Strictly speaking, this phase space would be a projection on the (x, x') plane of full phase space. In the simple situations in which no radiation, acceleration, or particle losses occur, the area in phase space is conserved under transformations represented by R. The latter statement is known as *Liouville's theorem*, which we will revisit in chapter 4. As a consequence of this theorem, the determinant of R is equal to one, i.e., $|R| = 1$.

The simplest case of a focusing function is the 'point' lens, a mathematical construct whereby $\kappa(s) = \delta(s)/f$. 'δ' represents the Dirac delta function and 'f' is a constant. By an integration of equation (1.1) with this focusing function, it is easy to see that 'f' represents the focal length of a zero-length lens located at $s = 0$. Figure 1.1 illustrates the zero-length or point lens.

By studying equation (1.2) for a point lens, we can identify the *principal trajectories* or rays illustrated in figure 1.1: for the first ray, $x'_1(0) = 0$, $x_1(0) = x_{10}$, leading to $x'_1(s) = -x_{10}(1/f)$, $x_1(s) = x_{10}(1 - s/f)$ for $s > 0$; for the second ray, $x_2(0) = 0$, $x'_2(0) = x'_{20}$, leading to $x_2(s) = x'_{20}s$. Thus, in this case, the matrix R is:

$$R_f = \begin{bmatrix} 1 & 0 \\ -1/f & 1 \end{bmatrix}. \tag{1.5}$$

Note that the first column in R represents ray 1 (coefficients multiplying initial conditions), while the second column represents ray 2.

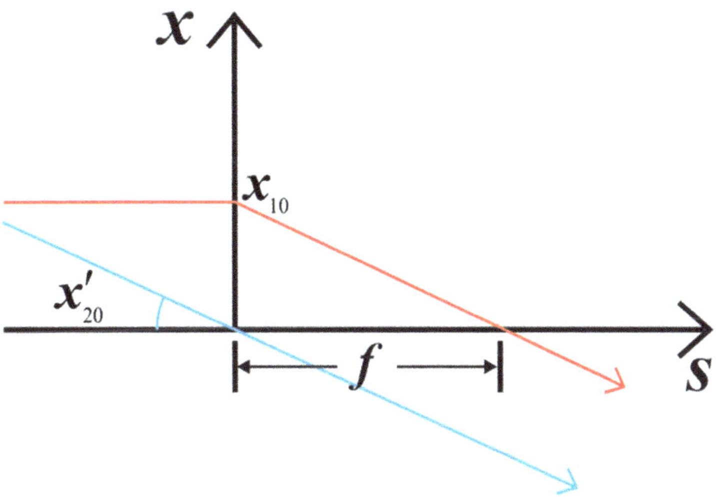

Figure 1.1. A 'point' lens at $s = 0$. The red and blue rays represent the principal trajectories or rays.

An equally important transfer matrix represents the ray transformation over a drift of length L:

$$R_{\mathrm{L}} = \begin{bmatrix} 1 & L \\ 0 & 1 \end{bmatrix}.$$ (1.6)

The matrices in equations (1.5) and (1.6) are the building blocks for transforming rays in linear systems with multiple lenses.

1.2 Thin lenses

If the extent of a real lens, as quantified by the *effective length* l_{eff} of the profile of the focusing function $\kappa(s)$, is much smaller than the focal length, then the lens is considered to be *thin*. (The point lens introduced above is an extreme example of a thin lens, because it remains so for *any $f > 0$*.) A useful construction that applies especially to lenses whose profiles $\kappa(s)$ are not dominated by *fringe fields*, i.e., extended 'tails,' is the *hard-edge model*. Figure 1.2 shows an example of an actual axial profile (of a magnetic field) related to $\kappa(s)$ and the corresponding hard-edge model.

The effective length of the hard-edge model is defined by

$$l_{\mathrm{eff}} \equiv \frac{1}{\kappa_{\mathrm{peak}}} \int_{-s_1}^{+s_1} \kappa(s)ds, \qquad \frac{1}{f} = l_{\mathrm{eff}}\kappa_{\mathrm{peak}}.$$ (1.7)

Therefore, the condition for a thin lens is $\kappa_{\mathrm{peak}} \ll 1/l_{\mathrm{eff}}^2$, or $f \gg l_{\mathrm{eff}}$. According to this criterion, a very strong lens can have a focal length comparable to its effective length, in which case the lens is *not* thin; or else the lens can be weak and have a very long focal length, satisfying the condition of a thin lens.

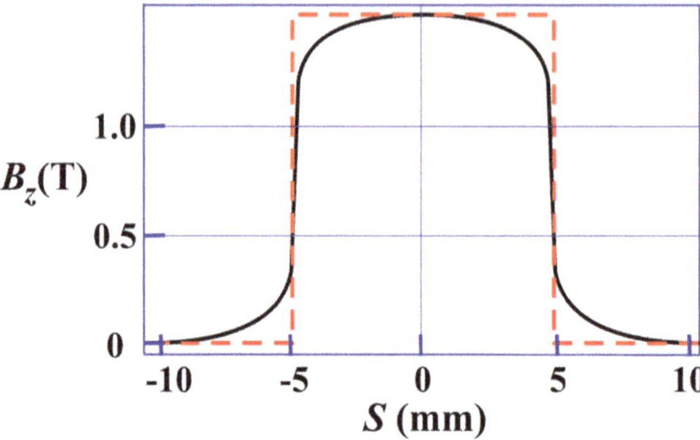

Figure 1.2. Actual axial profile of a magnetic field (related to the focusing function) and corresponding hard-edge model (red broken curve).

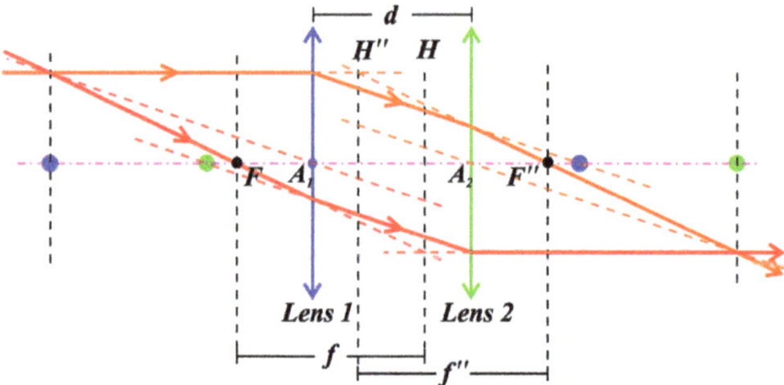

Figure 1.3. The system of two positive thin lenses is equivalent to a thick lens that has the principal planes H and H'' shown. Note that the focal lengths of the system are measured from the principal planes.

Ray tracing for a thin lens proceeds just as for the point lens (figure 1.1); so, for example, the exit x-coordinate of a given ray is unchanged after traversing the lens. Thus, the basic transfer matrix that connects ray coordinates on both sides of the lens is the same as that given in equation (1.5).

1.3 Thick lenses

Thick lenses in charged-particle optics can be treated just as thick lenses in standard optics. Concepts such as *principal planes, nodal points,* etc. also apply to magneto-static or electrostatic lenses [3]. The principal planes of a thick lens or a system of lenses are two hypothetical planes that connect locations of unit magnification, i.e., an object located on one plane is imaged on the second plane with the same size. Figure 1.3 illustrates the primary and secondary principal planes, H and H'', for a system of two focusing thin lenses separated by a distance d that have the same focal lengths. Ray tracing is done using the method of parallel rays as discussed in the standard optics book by Jenkins and White [4].

We can use equations (1.5) and (1.6) to find the matrix for the lens system shown in figure 1.3:

$$R = \begin{bmatrix} R_{11} & R_{12} \\ R_{21} & R_{22} \end{bmatrix} = \begin{bmatrix} 1 & 0 \\ -1/f_2 & 1 \end{bmatrix}\begin{bmatrix} 1 & d \\ 0 & 1 \end{bmatrix}\begin{bmatrix} 1 & 0 \\ -1/f_1 & 1 \end{bmatrix}, \tag{1.8}$$

or,

$$R = \begin{bmatrix} 1 - d/f_1 & d \\ -\left(\dfrac{1}{f_1} + \dfrac{1}{f_2} - \dfrac{d}{f_1 f_2}\right) & 1 - d/f_2 \end{bmatrix}. \tag{1.9}$$

By multiplying R with a drift matrix (length l_1) on the left and another (length l_2) on the right, we obtain a new matrix $\overline{R}$

$$\overline{R} = \begin{bmatrix} \overline{R}_{11} & \overline{R}_{12} \\ \overline{R}_{21} & \overline{R}_{22} \end{bmatrix} = \begin{bmatrix} R_{11} + l_2 R_{21} & R_{12} + l_1 R_{11} + l_2 R_{22} + l_1 l_2 R_{21} \\ R_{21} & R_{22} + l_1 R_{21} \end{bmatrix}. \tag{1.10}$$

Now, by setting $\overline{R}_{11} = 0$ we obtain the distance from the second lens to the second focal point of the system, F'' [5]:

$$l_2 = -\frac{R_{11}}{R_{21}} = \left(1 - \frac{d}{f_1}\right)\left(\frac{1}{f_1} + \frac{1}{f_2} - \frac{d}{f_1 f_2}\right)^{-1}. \tag{1.11}$$

The second factor on the right of equation (1.11) gives the inverse effective *object and image focal lengths f, f''* of the lens combination [4]:

$$\frac{1}{f} = \left(\frac{1}{f_1} + \frac{1}{f_2} - \frac{d}{f_1 f_2}\right) = \frac{1}{f''}. \tag{1.12}$$

The focal lengths f and f'' are *measured from the corresponding principal planes*, as shown in the figures.

Now assume that the second lens is divergent but that the two lenses have the same strength, i.e., the same focal lengths in magnitude: $f_1 = -f_2 > 0$. Then, from equation (1.12), the effective focal length is simply

$$f = \frac{f_1^2}{d} = \frac{f_2^2}{d}. \tag{1.13}$$

Furthermore, from equations (1.11) and (1.12), the second focal point of the system, F'', as measured from the second lens is

$$l_2 = f_1\left(\frac{f_1}{d} - 1\right). \tag{1.14}$$

This quantity is positive, indicating a convergent *equivalent thick lens* for the two-lens system if $f_1 > d$. Thus, we arrive at the important result that the combination of two lenses, one convergent (positive) and the other divergent (negative) that have the

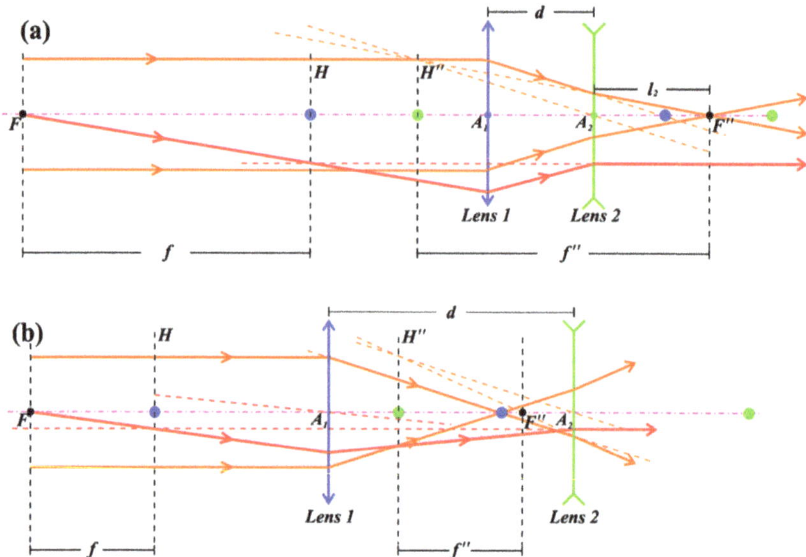

Figure 1.4. Combination of convergent (lens 1) and divergent (lens 2) lenses with the same strengths. In (a) the spacing d is *smaller* than the individual focal lengths ($f_1 = -f_2 > 0$); in (b) the spacing d is *larger* than the individual focal lengths. Note how the principal planes H and H'' are found using the parallel ray technique.

same strengths, leads to net focusing if their separation is smaller than the positive focal length f. This is relevant to the strong focusing scheme discussed in chapter 3 for particle accelerators.

Figure 1.4 shows the action of the focusing–defocusing lens combination for two values of their separation d. The geometrical construction that finds the principal planes uses the method of parallel rays mentioned above; examination of the figures may reveal the basic ideas. Note that the principal planes lie outside the space between the lenses in figure 1.4(a), unlike figure 1.3(a), for a pair of converging lenses with the same separation as in figure 1.4(a).

Additional relationships between the elements of the matrix in equation (1.10) and the important optical properties of a general lens system are discussed in an article by Halbach [6] and in appendix D of [5].

1.4 Transfer maps

The matrix formalism presented in the previous sections can be extended to higher orders. The complexity of the equations of motion to third order can be appreciated in Wiedemann's explicit treatment in his chapter 3 [2]; classic particle accelerator matrix software such as TRANSPORT [7], on the other hand, can perform only up to second-order calculations for most lattice elements. To third order, the ith component of the *six-dimensional* vector $\boldsymbol{\eta} = (x, x', y, y', z, \delta)^T$ (δ is the fractional momentum error—see section 6.5) is transformed between beamline planes '1' (exit) and '0' (input) according to [7]:

$$\eta_i(1) = \sum_j R_{ij}\eta_j(0) + \sum_{j,k} T_{ijk}\eta_j(0)\eta_k(0) + \sum_{j,k,l} U_{ijkl}\eta_j(0)\eta_k(0)\eta_l(0). \tag{1.15}$$

Most calculations in TRANSPORT employ up to the second-order matrix T. To perform second- and higher-order calculations, a more elegant and powerful (but also more demanding—from a mathematical viewpoint) approach is provided by Lie algebraic methods. The latter are based on the Lie algebra structure of Hamiltonian mechanics, as briefly discussed in appendix A and in far more detail by Dragt in [8]. Essentially, a beamline is described by a transfer *map* consisting of the product of the linear-transformation matrix R and the exponentials of *Lie operators* containing polynomials f_m [9]. The polynomials f_3, f_4, for example, describe second- and third-order effects, respectively.

Furthermore, since standard light optics can be shown to have a Hamiltonian structure [10], the same Lie algebra methods can be applied to both particle trajectories and light rays. In classical mechanics, particle trajectories obey the *principle of least action* [11]. Light rays, on the other hand, travel through optical systems such that their paths take the least times. This is known as *Fermat's principle* in geometrical optics [4].

To conclude this chapter, we present an example of non-paraxial imaging. We have used the MARYLIE code to image the letters in 'UMER' through a system of four focusing–defocusing quadrupole doublets; additional details are discussed in appendix A and in the MARYLIE manual [12]. Figure 1.5(a) shows the object to be imaged by a unit-magnification system employing 20 TeV protons and super-conducting magnets. Figure 1.5(b) displays the image obtained by tracing 2052 rays. The calculated transfer map differs slightly from a 6×6 unit matrix, which is enough to result in the blurring seen in figure 1.5(b). Most of the aberrations are caused by the quadrupole fringe fields; effects due to third-order distortion (see the letters 'U' and 'R'), astigmatism, and field curvature are present. Starting with a much smaller object and/or turning off the fringe fields considerably reduces the aberrations.

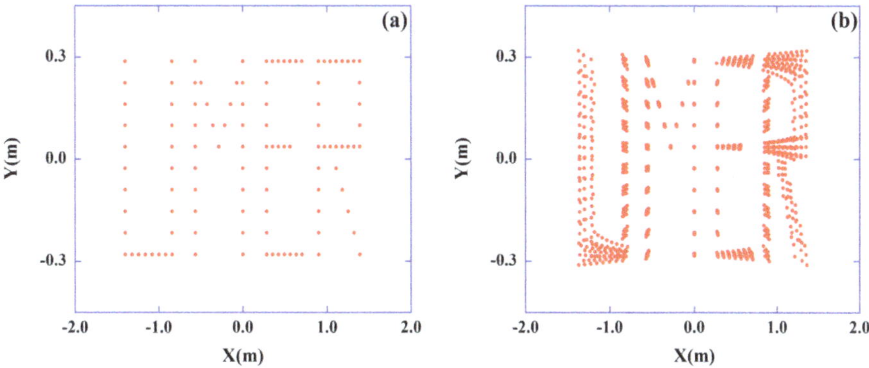

Figure 1.5. (a) The object to be imaged by four superconducting quadrupole doublets employing 20 TeV protons. (b) Image computed using the MARYLIE code; the effects of third-order aberrations are seen (see also appendix A and [12]).

1.5 Computer resources

Matrix multiplication, both symbolic and numeric, can be readily implemented in either *Mathcad* or *Matlab* or their freeware counterparts *SMath Studio* and *Octave* (see appendix A). Many standalone optics design programs are also available, but most are either too expensive or intended for professional use. Some software of this type is available as packages for codes such as *Mathematica* (e.g., Optics Lab).

We found the free program *Optgeo*, written by Jean-Marie Biansan, to be instructive and relatively easy to use (see appendix A for download information). The ray tracing in figures 1.3 and 1.4 can be reproduced without too much effort. We include two *Optgeo* files in this book's website. Alternatively, for *Mathcad* users we recommend a very well designed program for 2D optics called *2D Optical Ray Tracer*, which is written by Valter Kiisk of the University of Tartu (Estonia) and is available online for free. We include in our website a version with input that closely reproduces the results shown in figure 1.4(a). It is important to keep the lens thicknesses small relative to their focal lengths for comparison with figure 1.4, but we can also easily explore thick lens combinations. Figure 1.6 illustrates typical output.

For *Matlab*, there is at least one free app, *LensLab*, which can be used to reproduce the ray tracing of the lens doublets in this chapter. Another program that we found useful is *Ray Optics* for Android devices; there is a 'Pro' version that adds details of the calculations. *RayLab*, a free app for Apple's IOS devices, looks very professional and rather challenging as an educational tool for geometrical optics.

Thin magnetostatic lenses can be implemented in several of the most popular codes for particle accelerator design. (In fact, in some cases thin lenses *are* required for some operations, such as tracking in the code MAD.) Our website includes three example files for the program WINAGILE (see appendix A). The first file illustrates the implementation of one thin quadrupole lens; the other two files are examples of two thin quadrupole combinations. The files illustrate the concepts of this chapter and

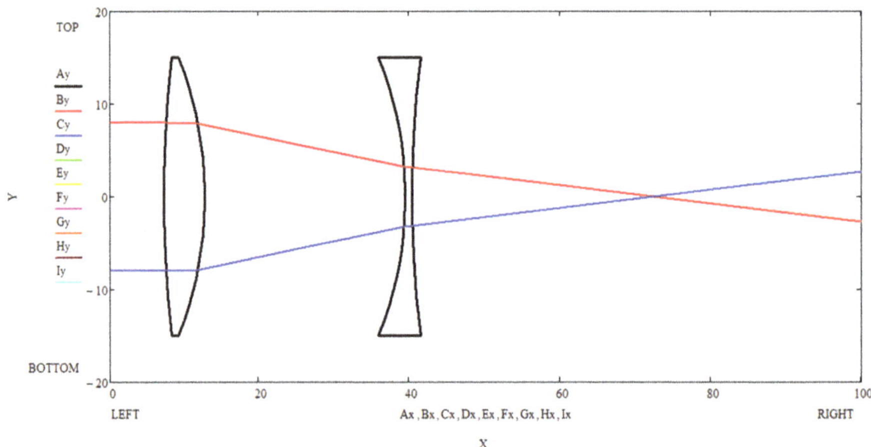

Figure 1.6. Graphics produced by the 2D optics *Mathcad* program *2D Optical Ray Tracer* by Valter Kiisk. Distances are in mm.

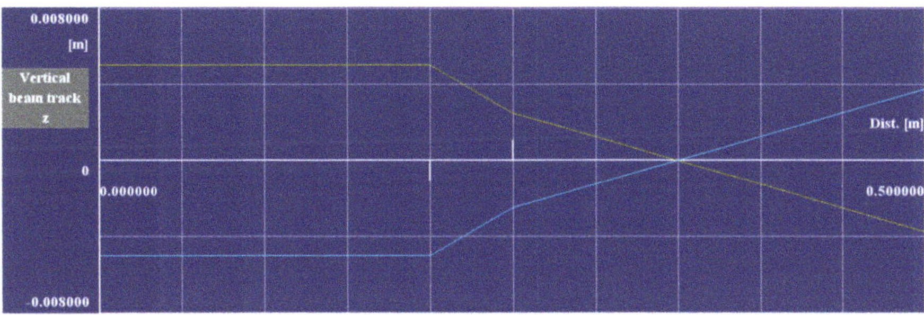

Figure 1.7. Trajectory tracking in WINAGILE for a combination of two thin quadrupoles of opposite strengths. The quadrupoles are located at $s = 20$ cm and 25 cm, while their individual focal lengths are $f_1 = -f_2 = 10$ cm. See also chapters 2 and 3 for additional important considerations.

serve as a preamble to quadrupole focusing (chapter 2) and to alternating gradient, or strong focusing (chapter 3). In appendix A, we indicate how to download, install, and run WINAGILE. Figure 1.7 is a screenshot from WINAGILE that shows trajectory tracking with a two thin quadrupole combination.

The example of non-paraxial imaging illustrated in figure 1.5 relies on the MARYLIE code [12] briefly described in section 1.4. Additional details pertaining to this code and the example can be found in appendix A.

References

[1] Steffen K G 1965 *High Energy Beam Optics,* (New York: Interscience Publishers/Wiley)

[2] Wiedemann H 2007 *Particle Accelerator Physics* 3rd edn (Berlin: Springer)

[3] Wollnik H 1987 *Optics of Charged Particles* (New York: Academic)

[4] Jenkins F A and White H E 1981 *Fundamentals of Optics* 4th ed (New York, NY: McGraw-Hill) 5

[5] Conte M and MacKay W W 1991 *An Introduction to the Physics of Particle Accelerators* (Singapore: World Scientific)

[6] Halbach K 1964 Matrix Representation of Gaussian Optics *Am. J. Phys.* **32** 90–108

[7] Carey D C, Brown K L and Rothacker F 1995 Third-order transport—a computer program for designing charged particle beam transport systems, SLAC-R-95-462, fermilab-Pub-95/068, UC-414

[8] Dragt, A J 1982 Lectures on nonlinear orbit dynamics *AIP Conf. Proc.* 87

[9] Dragt A J 2013 An overview of lie methods for accelerator physics *Proc. of PAC* (Pasadena, CA) 1134

[10] Dragt A J 1982 Lie algebraic theory of geometrical optics and optical aberrations *J. Opt. Soc. Am.* **72** 372–379

[11] Goldstein H 1980 *Classical Mechanics* 2nd edn (Reading, MA: Addison-Wesley)

[12] Dragt A *et al* 2003 *MARYLIE 3.0 Users' Manual: A Program for Charged Particle Beam Transport based on Lie Algebraic Methods* (College Park: University of Maryland) http://www.physics.umd.edu/dsat/docs/MaryLieMan.pdf

IOP Publishing

A Practical Introduction to Beam Physics and Particle Accelerators (Third Edition)

Santiago Bernal

Chapter 2

Linear magnetic lenses and deflectors

Charged-particle beams in accelerators can be focused and guided by means of electrical or magnetic elements. We restrict the discussion in this chapter to magnetic elements, because they are far more common in high-energy accelerators, but their electrical counterparts have essentially the same mathematical treatment. In some cases, the same magnet provides both deflection and focusing (a combined-function magnet), but most modern machines employ separate-function magnets. In other words, beam deflection and focusing are, ideally, completely independent of each other. In this chapter we discuss solenoids, magnetic quadrupoles, and dipole deflectors. Solenoids are not commonly employed as main lattice elements, but mostly to manipulate electron bunches near the source and for other specialized applications. Solenoid focusing per se is an interesting but non-straightforward concept, which we illustrate in detail and contrast with the idealized particle motion in a uniform B field. Magnetic quadrupoles, on the other hand, are the most widely used lattice elements in accelerators; they ideally provide *linear* strong focusing when paired in the alternating-gradient scheme (chapter 3). Naturally, important *nonlinear* focusing elements (e.g., sextupoles and octupoles) are also part of many accelerators, but we will defer discussing them to appendix B. We also present the Kerst–Serber equations and weak focusing in section 2.4. In section 2.5, we discuss dipole magnets and edge focusing, and in the last section, we describe computer resources for the chapter.

2.1 Magnetic rigidity, momentum, and cyclotron frequency

A charged particle moving in the presence of a uniform B field follows a circular trajectory. By equating the Lorentz force, $\mathbf{F} = q\mathbf{v} \times \mathbf{B}$, for a particle of charge q and velocity $v = \beta c\,\hat{\boldsymbol{\theta}}$ (c = speed of light in vacuum, $\hat{\boldsymbol{\theta}}$ = unit vector in the azimuthal direction), to the centripetal force, we obtain

$$qvB = \frac{\gamma m v^2}{\rho},\tag{2.1}$$

doi:10.1088/978-0-7503-4039-7ch2

where ρ is the radius of the circular orbit, m is the rest mass, and γ is the relativistic mass factor $\gamma = (1 - \beta^2)^{-1}$. Therefore, with the relativistic linear momentum defined as $p = \gamma m v$, we have

$$B\rho = \frac{p}{q}. \tag{2.2}$$

The quantity '$B\rho$' is called *magnetic rigidity*, and its units are Tm (Tesla-meter in SI units), or Gcm (Gauss-cm in centimeter–gram–second (CGS) units). The reason for this name is that particles with higher energies and therefore higher '$B\rho$' are harder to 'bend': a higher field would be required for a fixed orbital radius ρ. It is strange (and sometimes confusing) to give the explicit product of two quantities a single name, but this is almost universal usage in the accelerator-related literature. Perhaps only Wollnik [1] uses a single symbol (χ_B) for magnetic rigidity.

According to the special theory of relativity, the relativistic momentum, total energy E, and rest mass energy $E_0 = mc^2$ are connected by the relation

$$E^2 = p^2 c^2 + E_0^2 \approx p^2 c^2. \tag{2.3}$$

The last equality is valid for highly relativistic particles, i.e., particles for which $E \gg E_0$. For reference, $E_0 = 0.511$ MeV for the electron, and $E_0 = 938$ MeV, or 0.938 GeV, for the proton.

As an example, if B is given in units of Tesla (T) and ρ in meters, we can write the following relation for *relativistic electrons*:

$$B[\text{T}]\rho[\text{m}] = 0.3p[\text{GeVc}^{-1}]c, \tag{2.4}$$

where the momentum p is given in units of GeV c^{-1} and c is the speed of light. Thus, a 300 MeV electron moves with a circular trajectory that has a 1 m radius in a 1 T B field. As an exercise, the reader can write the corresponding relation for the magnetic rigidity of relativistic protons.

Finally, by writing $v = \rho(d\theta/dt) = \rho\omega_C$, we obtain from equation (2.1) the expression for the angular frequency of revolution of a particle of charge q in a uniform field B_0, which is called the *cyclotron frequency*:

$$\omega_C = \frac{qB_0}{\gamma m}. \tag{2.5}$$

A related frequency, which we will use later, is the *Larmor frequency*:

$$\omega_L = \frac{\omega_C}{2}. \tag{2.6}$$

2.2 Solenoid focusing

The solenoids used in particle accelerators are built from coils of normal conducting or superconducting wires and fed with very high direct currents to produce central B fields of up to several tesla (T). (For reference, 1 T = 10 000 G, and the magnitude of the Earth's B-field is approximately 0.5 G.)

From the discussion in the previous section, we can understand that charged particles in the *central* region of a solenoid of field $\boldsymbol{B}$ follow spiral trajectories that enclose the B-field lines. This motion in itself does not constitute focusing; the focusing effect occurs at the edges of the solenoid where the B-field has a radial component. In those fringe regions, the Lorentz force produces an azimuthal acceleration which in turn leads to *radially symmetric* focusing. To see this in detail, consider the passage of a particle through the fringe field region depicted in figure 2.1. Let us assume that the particle starts moving in a region at $s = s_0$ with $B = 0$, and with a velocity parallel to the solenoid axis, i.e., with $\dot{\theta}_0 = 0$, but with an offset radius ρ_0 near $r = 0$.

Therefore, the initial *canonical angular momentum* p_θ is zero, or, explicitly

$$\gamma m r^2 \dot{\theta} + q r A_\theta = 0, \tag{2.7}$$

where $A_\theta(r, z) = rB(r = 0, z)/2 \equiv rB_0(z)/2$ is the component of the vector potential that leads to a *rotationally symmetric* field $\overrightarrow{B}(r, z) = B_r\hat{\rho} + B_z\hat{z}$ (recall that $B = \nabla \times A$). The rotational frequency of particle motion inside the fringe region is then

$$\dot{\theta}(z) = -\frac{q}{\gamma m r}A_\theta(z) = -\frac{1}{2}\frac{qB_0(z)}{\gamma m} = \mp\omega_L(z), \tag{2.8}$$

according to equations (2.7), (2.5), and (2.6). The sign in front of ω_L in equation (2.8) is negative if q and B have the same signs and positive otherwise.

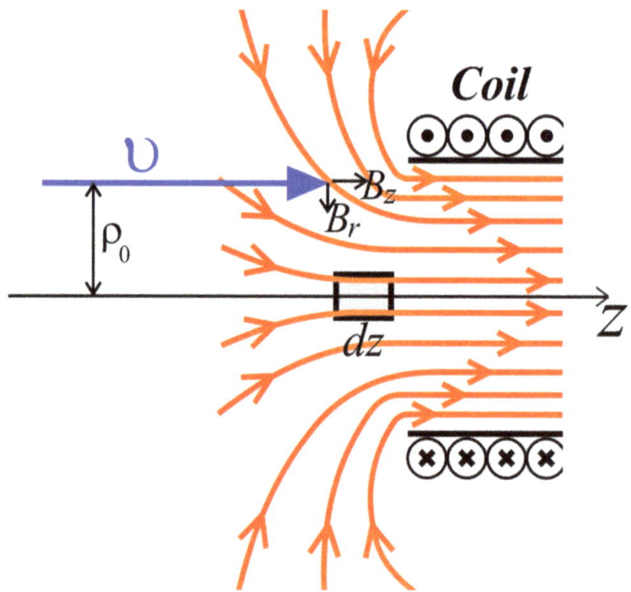

Figure 2.1. Geometry of field lines in the fringe region of a solenoid lens. The application of Gauss' law to the pillbox near the axis provides a way to show that $B_r \cong -(r/2)(\partial B_z/\partial z)_{r=0}$.

The radius of gyration can be found from the change in azimuthal or transverse linear momentum, $\Delta p_\perp$, not to be confused with p_θ, in the transition from '$B = 0$ space' into the fringe region:

$$\Delta p_\perp = q \int_0^{t_1} v_z B_r(z)dt = q \int_0^{z_1} B_r(z)dz, \tag{2.9}$$

which can be further evaluated using $B_r(z) \cong -(r/2)(\partial B_z/\partial z)_{r=0}$. The latter result follows from the assumption of a linear radial fringe field $B_r(z) = B_0(z)r$ near $r = 0$, rotational symmetry, and $\nabla \cdot B = 0$ in cylindrical coordinates. (Alternatively, we can use Gauss' Law and the pillbox indicated in figure 2.2. See the article by McDonald [2].) We obtain $\Delta p_\perp = -q\rho_0 B_0/2$, which implies a radius of gyration equal to $\rho_0/2$ (see equation (2.2)). Furthermore, the opposite change in transverse linear momentum occurs in the transition from the region inside the solenoid to the region of '$B = 0$ space' at the opposite end of the solenoid.

In conclusion, particles initially moving parallel to and near the axis in a solenoid rotate with the *local* Larmor frequency and with a radius equal to one-half of the initial offset; this implies that the particles periodically cross the *solenoid axis* at $r = 0$. It is important to understand that particles rotate around the centers of their individual orbits at the cyclotron frequency, but the rotations around the solenoid axis occur at the Larmor frequency. The geometry of these rotations is illustrated in figure 2.2.

It is convenient to describe the motion in a reference frame that rotates with the Larmor frequency $\dot{\theta}$, i.e., equation (2.8). It is clear that the particles follow *simple*

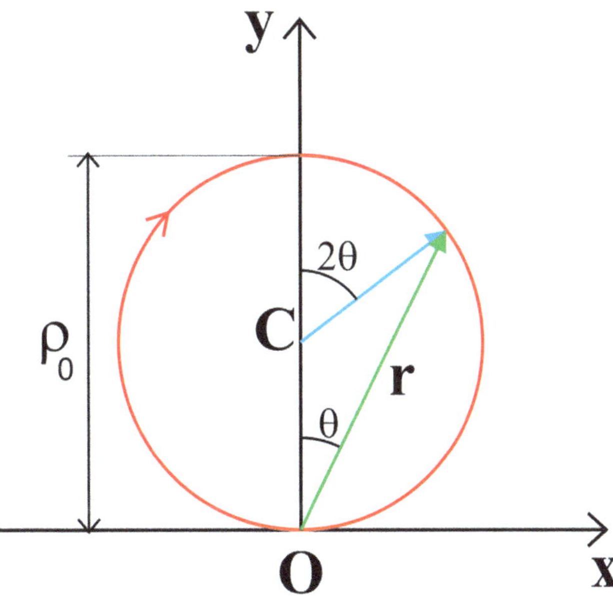

Figure 2.2. Orbital geometry in solenoid focusing of particles with an initial offset equal to ρ_0 and no canonical angular momentum. 'O' is on the axis of the solenoid and 'C' is the center of the individual orbit.

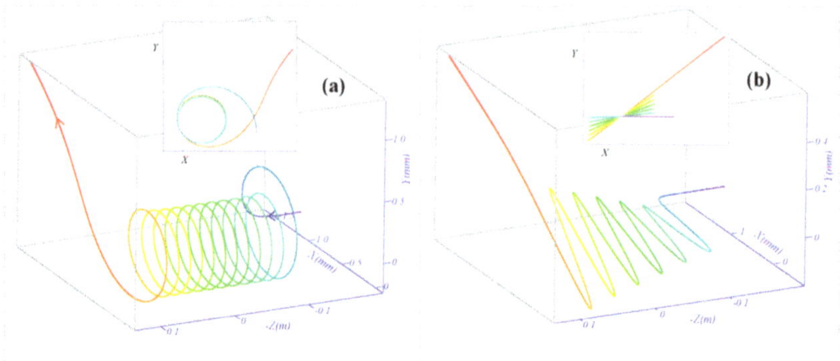

Figure 2.3. Electron trajectories in a solenoid lens as seen in (a) the lab frame, and (b) the Larmor frame. The insets show the projections on the transverse (X, Y) planes. See also the section on computer resources (2.6).

harmonic motion in both transverse coordinates. Thus, the wave number along the X or Y-axis in the *Larmor frame* is

$$k_{X,\,Y}(z) = \frac{\omega_L(z)}{v_z} = \frac{1}{2}\frac{B(z)}{(B\rho)}. \tag{2.10}$$

Finally, using equation (2.10) and the paraxial equation of motion (1.1), the transverse focusing function for a solenoid can be written as

$$\kappa_S(s) = \frac{1}{4}\frac{B^2(s)}{(B\rho)^2}, \tag{2.11}$$

where $B\rho$ is the magnetic rigidity (which was defined in equation (2.2)), and we have reverted to using 's' instead of 'z' for the axial coordinate. Note that solenoid focusing is always positive. Figure 2.3 depicts the trajectories of a 10 keV electron beam in a 0.14 T (peak axial B-field), 0.19 cm (effective length) solenoid in both the laboratory and Larmor frames (see also the section on computer resources (2.6)). Additional insights into solenoid focusing can be found in an *American Journal of Physics* article by Vinit Kumar [3].

2.3 Quadrupole focusing

In the absence of electrical currents in the region traversed by a beam inside a magnetic lens, the B-field can be derived from a scalar potential: $\vec{B} = -\nabla\Phi$. In addition, if the lens is long compared to its aperture, we can ignore any dependence of Φ on the axial coordinate, z. We have, in effect, a 2D problem and the scalar potential in cylindrical coordinates can be written as [4]

$$\Phi(r,\,\theta) = \sum_{n=1}^{\infty}\Phi_n = \sum_{n=1}^{\infty}[c_n r^n \cos(n\theta) - d_n r^n \sin(n\theta)], \tag{2.12}$$

which is applicable inside the magnet. (An equivalent, alternative formulation in terms of the expansion of a complex B-field is presented in appendix B.) The term n

= 1 corresponds to a linear potential, i.e., a constant (dipole) field. The term $n = 2$ yields a quadratic (in r) potential

$$\Phi_2(r, \theta) = c_2 r^2 \cos(2\theta) - d_2 r^2 \sin(2\theta). \tag{2.13}$$

The part of the potential proportional to d_2 is called the *normal quadrupole component*, while the part proportional to c_2 is the *skew quadrupole component*. The normal component has the form

$$\Phi_{\text{Normal}} = -2d_2xy \tag{2.14}$$

in Cartesian coordinates. Therefore, the x-component of the Lorentz force on a particle initially moving with velocity v along the $+z$-axis is

$$F_x = -qvB_y = -2qvd_2x. \tag{2.15}$$

Similarly, the y-component of the force is

$$F_y = qvB_x = 2qvd_2y. \tag{2.16}$$

We can see from equations (2.15) and (2.16) that an ideal normal quadrupole provides *linear* focusing in the horizontal plane and *linear* defocusing in the vertical plane. This latter statement assumes that the product qd_2 is positive and that v is in the $+z$ direction. Figure 2.4 illustrates the B field geometry and the forces in such a situation.

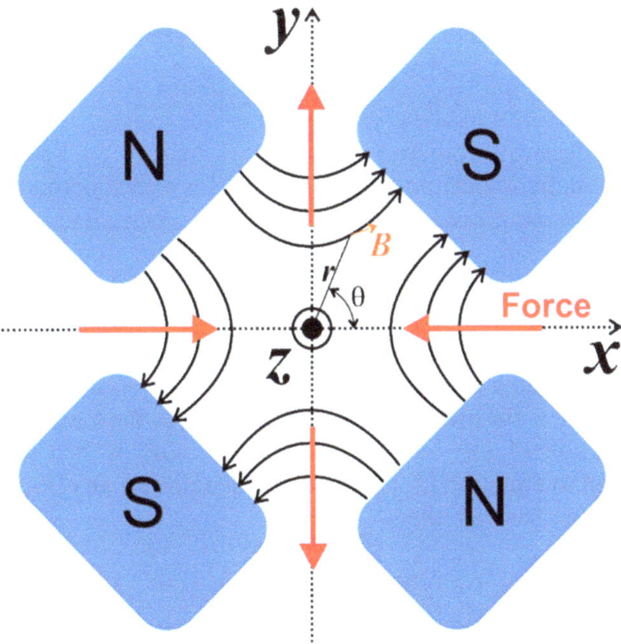

Figure 2.4. Quadrupole magnet for horizontal focusing of beams of positively charged particles. The red arrows indicate the magnetic forces acting on particles moving out of the plane of the figure.

To make the connection with equation (1.1) and find the form of the focusing function $\kappa(s)$, we first realize that the time and axial derivatives of x are related through

$$\dot{x} = v x', \quad \ddot{x} = v^2 x'', \tag{2.17}$$

so Newton's second law yields

$$x'' = -\frac{2d_2}{(B\rho)} x = -\frac{g_x}{(B\rho)} x, \tag{2.18}$$

and similarly for the y-direction, except for the sign on the right-hand-side. The *B-field gradients* in the x and y directions (see equations (2.15) and (2.16)) are defined by

$$g_x = \frac{\partial B_y}{\partial x}, \quad g_y = \frac{\partial B_x}{\partial y}. \tag{2.19}$$

We generalize and include an *axial dependence* for the field gradients and write

$$\begin{aligned} x''(s) + \kappa_x(s)x(s) &= 0, \\ y''(s) - \kappa_x(s)y(s) &= 0, \end{aligned} \tag{2.20}$$

where

$$\kappa_x(s) = \frac{1}{(B\rho)} g_x(s). \tag{2.21}$$

The magnitude of the focal length is then related to the *integrated gradient* by:

$$\frac{1}{f} = \frac{1}{(B\rho)} \int g(s)ds = \frac{1}{(B\rho)} g_0 l_{\text{eff}}, \tag{2.22}$$

where l_{eff} is the effective length and g_0 is the peak on-axis gradient.

Typically, magnetic quadrupole lenses, based on either permanent magnets or electromagnets, are employed to focus relativistic particles, while electrostatic quadrupoles (based on electric field gradients) are used at low energies. Quadrupole focusing is termed 'strong' focusing because it is more efficient than focusing based on fringe effects (e.g., solenoid focusing or focusing performed by cyclotron magnets). We can demonstrate this by comparing the focal lengths of a quadrupole doublet and a solenoid when both have the same peak *B-fields* and the same overall lengths. If the quadrupoles have aperture a, the magnitude of their peak fields is $g_0 a = B_0$ (ideal linear quadrupoles); furthermore, if their separation is $2l_{\text{eff}}$, with l_{eff} as defined in equation (2.22), and the solenoid has an effective length equal to $2l_{\text{eff}}$ and the same field B_0, then

$$\frac{f_{\text{solenoid}}}{f_{\text{quad doublet}}} \approx \frac{4l_{\text{eff}}^2}{a^2}, \tag{2.23}$$

as the reader can verify from equations (1.13) (chapter 1), (2.11), and (2.22). Since ordinarily $a \ll l_{\text{eff}}$, the doublet is significantly stronger than the equivalent solenoid.

2.4 The Kerst–Serber equations and weak focusing

The discussion in section 2.1 of particle motion in a uniform B-field led to the formula for the cyclotron frequency, equation (2.5), which can be complemented with the cyclotron radius or radius of the *design orbit* ρ_0:

$$\rho_0 = \frac{\gamma m v}{q B_0} = \frac{(B\rho)}{B_0}. \tag{2.24}$$

Although equation (2.24) may seem tautological, it must be understood that specifying the magnetic rigidity of a particle does not automatically yield either the actual design orbit radius or the B-field.

An important question in the context of circular accelerators is the *stability* of the orbits. Using cylindrical coordinates (r,θ,z), the radial equation of motion is $\gamma m \ddot{r} - \gamma m r \dot{\theta}^2 = q r \dot{\theta} B_z$. We recover equation (2.24) if $r = \rho_0$, $B_z = B_0$, both constant, and $r\dot{\theta} = -v$. The latter choice of sign indicates that the unit vectors $\hat{r}$, $\hat{\theta}$, $\hat{z}$ form a right-handed coordinate system in which θ increases in the counterclockwise direction, so a positively charged particle moves in a clockwise direction if the B-field is pointing in the $+z$ direction. Figure 2.5 illustrates the geometry of the problem.

Let us now assume that the particle moves away from the design orbit with a small offset x, so that $r = \rho_0 + x$, with $x \ll \rho_0$. Furthermore, let us expand the vertical B-field in x and keep only the linear term:

$$B_z(r, x) = B_0 + x\left(\frac{\partial B_z}{\partial r}\right)_{r=\rho_0} = B_0\left(1 - n\frac{x}{\rho_0}\right), \tag{2.25}$$

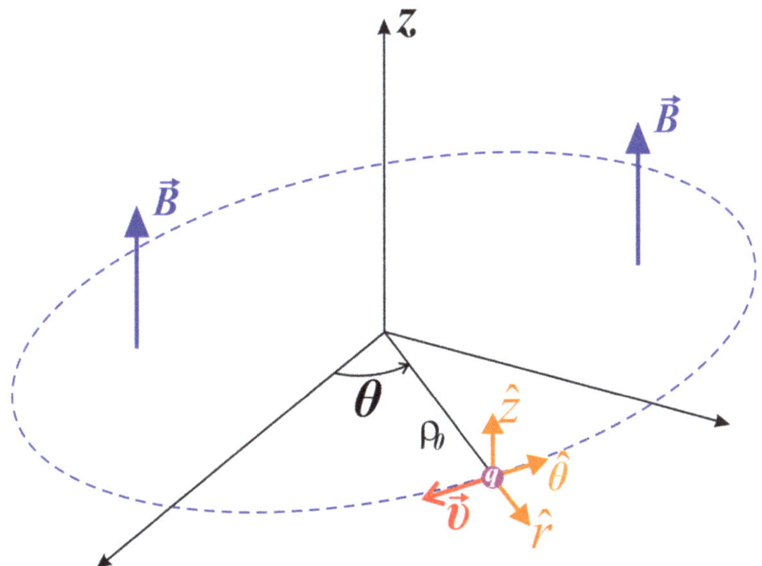

Figure 2.5. Reference orbit and cylindrical coordinate system. The particle is positively charged.

where we have introduced the *field index n* defined by

$$n \equiv -\frac{\rho_0}{B_0}\left(\frac{\partial B_z}{\partial r}\right)_{r=\rho_0}. \tag{2.26}$$

With these definitions, an equation of motion for x can easily be derived [5, 6]:

$$x''(s) + \frac{1}{\rho_0^2}(1-n)x(s) = 0. \tag{2.27}$$

Similarly, for the vertical motion we have

$$z''(s) + \frac{n}{\rho_0^2}z(s) = 0. \tag{2.28}$$

Equations (2.27) and (2.28) are called the Kerst–Serber equations after D Kerst and R Serber, who in 1941 studied the stability of transverse motion in a betatron accelerator. The equations indicate that particles undergo harmonic oscillations around the design orbit, both radially and in the vertical plane, but with different frequencies. These oscillations are called *betatron oscillations*. Furthermore, this motion is stable if the associated focusing function (see equation (1.1)) follows $\kappa_{x,z} > 0$, i.e., for $0 < n < 1$.

In the radial direction (see equation (2.27)), the focusing function has two parts, one proportional to $1/\rho_0^2$ and the other proportional to $-n$. The first part is a geometrical effect that corresponds to what is known as *weak focusing*, while the second part arises because of the appearance of a radial component of the B-field that varies linearly with the height z over the orbital plane. This latter effect can be seen from $\nabla \times \mathbf{B} = 0$, which implies $\partial B_r/\partial z = \partial B_z/\partial r$ or $B_r = -(B_0/R_0)nz$. In the vertical plane there is no geometrical focusing, but a focusing function proportional to $+n$ (see equation (2.28)), the opposite of the situation in the horizontal plane. We can recast equation (2.27) to see the quadrupole term more clearly (see equation (2.19)):

$$x''(s) + \left(\frac{1}{\rho_0^2} + \frac{g_x}{(B\rho)}\right)x(s) = 0, \quad g_x = \left(\frac{\partial B_z}{\partial x}\right)_{x=0}, \tag{2.29}$$

since $B_0\rho_0 = B\rho$, and we are using 'z' instead of 'y' for the vertical coordinate.

Furthermore, from equation (2.27), the wave number associated with the radial motion is $k_x = \sqrt{\kappa_x} = (1-n)^{1/2}/\rho_0$. Thus, the number of radial oscillations per turn, which defines the *radial betatron tune*, ν_x, is

$$\nu_x = \frac{2\pi\rho_0}{\lambda_x} = k_x\rho_0 = (1-n)^{1/2}. \tag{2.30}$$

Similarly, for the *vertical betatron tune*:

$$\nu_z = k_z\rho_0 = n^{1/2}. \tag{2.31}$$

We see that with weak focusing, the betatron tunes, as the field indices, cannot be larger than one if the orbits are to be stable. This limitation is overcome in a different scheme, strong focusing, which is discussed in chapter 3.

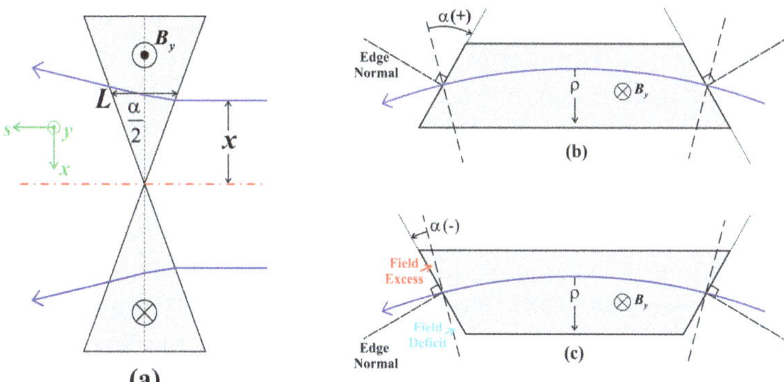

Figure 2.6. Edge focusing of a positively charged particle: (a) thin magnetic wedge, (b) magnet with horizontal edge defocusing, and (c) magnet with horizontal edge focusing. The blue traces in (b) and (c) indicate the reference trajectories (radius ρ).

2.5 Dipoles and edge focusing

Let us assume that a charged particle of magnetic rigidity $B\rho$ moves through a region of uniform field ΔB and length L. By equating the centripetal force to the Lorentz force and recalling that $\dot{\theta} = \theta' v$ and the definition of magnetic rigidity, we find the deflection angle $\Delta\theta$ caused by this dipole field:

$$\Delta\theta = \frac{\Delta B \cdot L}{(B\rho)}. \tag{2.32}$$

(Alternatively, we can simply use $\Delta\theta = \Delta s/\rho$, where Δs is the arc length, and multiply by $\Delta B/B$ with the understanding that $\Delta B = B$ for a uniform field.)

More generally, we can write

$$\Delta\theta = \frac{1}{(B\rho)} \int_{s_1}^{s_2} B(s)\,ds, \tag{2.33}$$

for the deflection through a non-uniform field $B(s)$. Note that we can obtain equation (2.33) by setting $ds = \rho(s)d\theta$, and using $B(s)\rho(s) = B\rho$, for all s, the latter equality being valid without acceleration (other than centripetal). Here, we see the potential for confusion, because we need to distinguish $B(s)$ in the integrand in equation (2.33) from B in the factor outside the integral. The latter B is not to be taken separately, because the actual factor is $B\rho = p/q$, a constant. Note also that the net deflection is proportional to the 'integral Bds,' regardless of the details of how the field varies along the trajectory through the dipole magnet.

Motion through the boundary regions in a real dipole magnet leads to focusing effects. In essence, a particle moving with a (transverse) offset distance from the reference trajectory can see either a 'deficit' or an 'excess' of integrated transverse magnetic field. Equivalently, according to equation (2.32), particles are deflected by an angle proportional to the traversed distance 'L' in the boundary region, which in turn is to first order proportional to the transverse offset. We can see this explicitly with the help of figure 2.6(a), which represents the effect of either edge in the dipole magnet depicted in figure 2.6(b).

Figures 2.6(b) and 2.6(c) show reference trajectories forming an angle $|\alpha|$ with the normal to the pole faces. The intermediate case between figures 2.6(b) and 2.6(c) is not shown, but we can deduce that it corresponds to $\alpha = 0$: the reference trajectory enters and exits the pole faces along the normal to the faces. This latter magnet is called a *sector magnet*. Thus, the thin wedge magnet of figure 2.6(a) is designed so that it *subtracts* to the field of an ideal sector magnet on the upper part (of the reference trajectory, at the boundary), but adds to the field on the lower part. In other words, the magnet of figure 2.6(b) is equivalent to the superposition of a sector magnet and a thin wedge magnet (figure 2.6(a)) at each end (note the directions of the bending fields in figures 2.6(a) and (b)). Furthermore, the distance in the thin wedge of figure 2.6(a) is $L = 2x \tan(|\alpha|/2) \cong x \tan |\alpha|$. Therefore, using equation (2.32) we can write

$$\Delta\theta = \frac{B_y \tan \alpha}{(B\rho)} x. \tag{2.34}$$

By convention the angle α for the magnet in figure 2.6(b) is *positive*, so equation (2.34) implies that net defocusing results (i.e., a positive slope for particles moving from right to left). In contrast, the magnet in figure 2.6(c) has a negative α and performs focusing. To summarize, edge focusing is represented by the matrices

$$R_{xx\text{[wedge]}} = \begin{bmatrix} 1 & 0 \\ +\dfrac{\tan \alpha}{\rho} & 1 \end{bmatrix}, \quad R_{yy\text{[wedge]}} = \begin{bmatrix} 1 & 0 \\ -\dfrac{\tan \alpha}{\rho} & 1 \end{bmatrix}. \tag{2.35}$$

R_{xx} and R_{yy} represent thin lenses for the horizontal and vertical planes, respectively.

Although no edge focusing occurs for a sector magnet, the magnet provides net focusing in the horizontal plane. This is just a geometric effect that corresponds to the weak focusing discussed in connection with equation (2.27); the focusing constant is then $\kappa = 1/\rho_0 = 1/\rho$. In the vertical plane, the sector magnet is represented by a drift. In matrix form (see equation (3.9), chapter 3), a horizontal sector magnet is represented by the matrices

$$R_{xx\text{[sector]}} = \begin{bmatrix} \cos\theta & \rho\sin\theta \\ -\dfrac{\sin\theta}{\rho} & \cos\theta \end{bmatrix}, \quad R_{yy\text{[sector]}} = \begin{bmatrix} 1 & \rho\theta \\ 0 & 1 \end{bmatrix}. \tag{2.36}$$

If $|\alpha|$ in a magnet is one-half of the bending angle θ, we have a *rectangular magnet*. It can be shown by matrix multiplication that in this case, the geometric and edge effects exactly balance each other out, yielding no net focusing in the horizontal plane. In the vertical plane (perpendicular to the bending plane), however, the rectangular magnet can be shown to introduce edge *focusing*.

The section on computer resources below provides exercises that illustrate edge focusing and sector and rectangular magnets.

2.6 Computer resources

Particle motion through the UMER short solenoid whose field profile is given by equation (B.21) in appendix B can be studied using the *Mathcad* program

TrajShortUMERSolenoid.xmcd. The trajectories shown in figures 2.3(a) and (b) can be reproduced by another program, **TrajExtSolenoid.xmcd**. In this case, the solenoid's on-axis B_Z-field profile is defined as in the Halbach model described in [7]. Furthermore, calculations can be conducted to first, third, or fifth orders in r. More interestingly, the results show that the simple model of motion through the fringe region of a solenoid as discussed in section 2.2 is not followed exactly; because the transition from 'zero' to the peak fields in the more realistic models is smooth, the rotational frequency varies accordingly. Thus, the Larmor frequency, defined as half the peak cyclotron frequency, is but one, or close to one, frequency component of the Fourier spectrum of the rotations. Motion in the Larmor frame, however, can be seen to resemble simple harmonic motion.

Particle trajectories in a sector magnet (section 2.5) can be studied numerically using the *Mathcad* program **3_Lorentz_Bending_Magnet.xmcd**, originally developed at the University of Colorado, Boulder, and available through this book's website. Note that $q = 1$, $m = 1$ and the special construction of the sector: the bending radius originates at the apex of the magnet. Thus, with the default values of the program, the inverse focal length (equation (2.36)) is just $1/f = \sin\theta/\rho = \sin 30^0 = 0.5$, since $\rho = 1$ (SI units implied).

The WINAGILE (see appendix A) files **S-bend.lat** and **R-bend.lat** contain single $10°$ bends and help us to understand the matrix representation and focusing properties of sector and rectangular magnets with 'flat' fields. Run either file as a 'transfer line' and click on the last row to reveal the matrix elements. Figure 2.7, top,

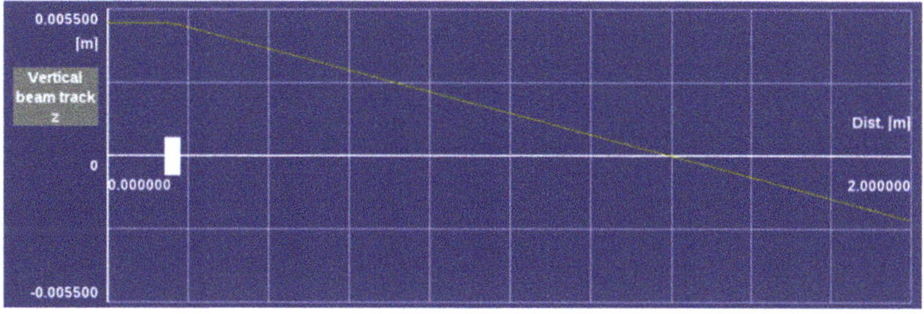

hh11	hh12	hv11	hv12	hs11	hs12	
1.000000	1.999809	0.000000	0.000000	0.000000	0.321942	x
hh21	hh22	hv21	hv22	hs21	hs22	
0.000000	1.000000	0.000000	0.000000	0.000000	0.174977	dx/ds
vh11	vh12	w11	w12	vs11	vs12	
0.000000	0.000000	-0.483183	1.762765	0.000000	0.000000	z
vh21	vh22	w21	w22	vs21	vs22	
0.000000	0.000000	-0.806014	0.870921	0.000000	0.000000	dz/ds
sh11	sh12	sv11	sv12	ss11	ss12	
0.000000	0.000000	0.000000	0.000000	0.000000	0.000000	s
sh21	sh22	sv21	sv22	ss21	ss22	
0.000000	0.000000	0.000000	0.000000	0.000000	0.000000	dp/p

Figure 2.7. Output produced by the WINAGILE file **R-bend.lat**. Top: matrix elements. Note in particular the element -vv21; it is the inverse of the focal length in the vertical plane. Particle tracking in the vertical plane is shown in the bottom graph.

shows the output that can be expected for a rectangular magnet. Furthermore, tracking using a particle starting at +0.005 m in the vertical plane illustrates vertical focusing (figure 2.7, bottom).

References

[1] Wollnik H 1987 *Optics of Charged Particles* (New York: Academic)

[2] McDonald K T 2011 Expansion of an axially symmetric, static magnetic field in terms of its axial field *Joseph Henry Laboratories* (Princeton, NJ: Princeton University) (available online)

[3] Kumar V 2009 Understanding the focusing of charged particle beams in a solenoid magnetic field *Am. J. Phys.* **77** 737

[4] Rosenzweig J B 2003 *Fundamentals of Beam Physics* (Oxford: Oxford University Press)

[5] Reiser M 2008 *Theory and Design of Charged Particle Beams* 2nd edn (Weinheim: Wiley-VCH)

[6] Bryant P J and Johnsen K 1993 *The Principles of Circular Accelerators and Storage Rings* (Cambridge: Cambridge University Press)

[7] Bernal S *et al* 2006 RMS envelope matching of electron beams from 'zero' current to extreme space charge in a fixed lattice of short magnets *Phys. Rev. ST Accel. Beams* **9** 064202

IOP Publishing

A Practical Introduction to Beam Physics and Particle Accelerators (Third Edition)

Santiago Bernal

Chapter 3

Periodic lattices and functions

A periodic lattice consists of a repeating focusing structure composed of one or more lenses, not all necessarily identical, whose function is to allow the transport of charged-particle beams over long distances. The repeating structure or cell constitutes one *lattice period*; the focusing function introduced in chapter 1 then becomes a periodic function. Furthermore, if focusing is *axisymmetric*, focusing of particles occurs with equal strengths (and signs) in both transverse planes; this is the case with a *solenoid lattice*, which requires a single focusing function. We consider a periodic linear solenoid lattice in the first section. Although it is not commonly used, the solenoid lattice provides a simple system that allows us to study stable beam transport and introduce the *phase advance per period* without space charge. In the second section, we analyze the *alternating-gradient (AG)* focusing system, which consists of quadrupoles paired with opposite gradient polarities. If the strengths of the quadrupoles are the same (except for the sign), the configuration is called symmetric FODO, which stands for 'Focusing–DefOcusing.' For various reasons, however, it is customary to power the two quadrupoles per FODO cell with different strengths; this case is also considered. In section 3.3, we summarize the classic *Courant–Snyder theory*, the 'bread-and-butter' theory of accelerator physics. In section 3.4, we introduce the *uniform-focusing approximation*, an idealized constant focusing function that is widely used for theoretical and computational studies. Section 3.5 covers *linear dispersion* and section 3.6 introduces additional important concepts that are especially applicable to rings. Finally, the last section covers computer resources and exercises.

doi:10.1088/978-0-7503-4039-7ch3

3.1 Solenoid lattice

Focusing by a single solenoid was discussed in section 2.2. The transfer matrix for a solenoid in the Larmor frame of reference can be written as

$$
\boldsymbol{R}_S = \begin{bmatrix} \cos\theta & \dfrac{l_S}{\theta}\sin\theta \\[2ex] -\dfrac{\theta}{l_S}\sin\theta & \cos\theta \end{bmatrix},
\tag{3.1}
$$

where $\theta = \sqrt{\kappa_S}\, l_S$, κ_S is given by equation (2.11), and l_S is the effective length of the solenoid. (We postpone the description of ray transformation in the laboratory frame, i.e., a transformation that includes rotation, to section 5.1.) It is implicit in equation (3.1) that $\kappa_S \equiv \kappa_S(s)$, i.e., the focusing function can vary with distance. Note that in the limit of small θ, the matrix $\boldsymbol{R}_S$ reduces to the thin-lens form given by equation (1.5). In that limit, $\cos\theta = 1$, $\sin\theta = \theta$, and $\kappa_S l_S = 1/f$.

Let us now consider N solenoids in a linear configuration with a spacing L between them. The matrix corresponding to *one period* of the solenoid lattice is then the product $\boldsymbol{R}_S\boldsymbol{R}_L$, where $\boldsymbol{R}_L$ is the matrix representing a drift (equation (1.6)). Let us write the one-period matrix in the form

$$
\boldsymbol{R}_1 = \boldsymbol{R}_S\boldsymbol{R}_L = \begin{bmatrix} A & B \\ C & D \end{bmatrix},
\tag{3.2}
$$

so that a ray $[r_{s+1},\, r'_{s+1}]$ at a plane labeled by '$s+1$' is related to a ray at plane 's' by

$$
\begin{aligned}
r_{s+1} &= A r_s + B r'_s, \\
r'_{s+1} &= C r_s + D r'_s.
\end{aligned}
\tag{3.3}
$$

A *difference equation* can be easily obtained from these relations:

$$
r_{s+2} - 2b r_{s+1} + r_s = 0,
\tag{3.4}
$$

where we have used $b = (A + D)/2 \equiv (1/2)Tr[\boldsymbol{R}_1]$ and $AD - BC = 1$. The symbol 'Tr' stands for 'trace,' while the last equality is equivalent to $\det[\boldsymbol{R}_1] = 1$ (see chapter 1 and section A.1 in appendix A). We now look for a *periodic ray solution* of the form $r_s = r_0 e^{isq}$ (naturally, only the real—or the imaginary—part of 'r' is physical). When r_s is substituted into equation (3.4), we find that $e^{iq} = b \pm i\sqrt{1 - b^2}$. By setting $b = \cos\sigma_0$, we see that σ_0 is real and the ray solution is confined if $|\,b\,| \leqslant 1$, i.e., if the *trace* of the one-period matrix satisfies

$$
|\,Tr[\boldsymbol{R}_1]| \leqslant 2.
\tag{3.5}
$$

The quantity σ_0 is the change in phase of the ray when it advances by one period; thus, it is called the *phase advance per period*. In practice, the phase advance σ_0 is specified, and the required magnet strength is obtained by solving the following equation for $\theta = \sqrt{\kappa_S}\, l_S$:

$$
\cos\sigma_0 - \frac{1}{2}Tr[\boldsymbol{R}_1] = 0.
\tag{3.6}
$$

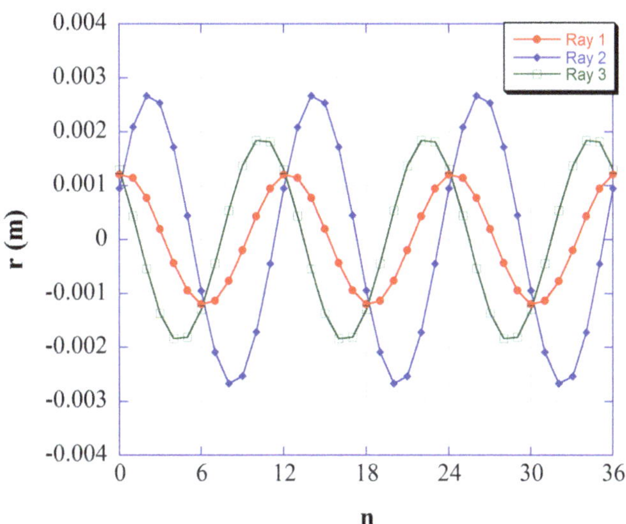

Figure 3.1. Ray propagation in a periodic lattice with 36 solenoids. Three rays with different initial conditions (r_0, r'_0) start at $n = 0$. The phase advance per period is $\sigma_0 = 30°$.

From the product of the solenoid and drift matrices, equations (3.1) and (1.6), we can identify $A = \cos\theta$, $B = L\cos\theta + (l_S/\theta)\sin\theta$, $C = -(\theta/l_S)\sin\theta$, and $D = \cos\theta - (L\theta/l_S)\sin\theta$, so that condition 3.5 can be explicitly written as

$$\frac{1}{2}\,|\,Tr[\mathbf{R}_1]| = \left|\cos\theta - \frac{L\theta}{2l_S}\sin\theta\right| \leqslant 1. \tag{3.7}$$

In the limit of thin lenses we have

$$\left|1 - \frac{L}{2f}\right| \leqslant 1, \tag{3.8}$$

from which the condition $L \leqslant 4f$ easily follows. In the next section, we will find that the condition $L \leqslant 2f$ applies to an alternating-gradient quadrupole lattice. Clearly, the combination of positive lenses, such as solenoids, allows for greater separation.

As an example, let us consider a periodic lattice of solenoids that have an effective length of $l_S = 0.065$ m and a drift space of $L = 0.10$ m. If we specify a phase advance per period of $\sigma_0 = 30°$, we obtain $\theta = 0.3272$ from equation (3.6), and $\kappa_S = 25.35$ m^{-2} (see the definitions below equation (3.1)). Ray propagation in this lattice is illustrated in figure 3.1.

3.2 FODO lattice

A *symmetrical* FODO lattice is realized with paired quadrupole lenses that have the same strength but opposite gradient polarities. Figure 3.2 is a schematic representation of the focusing function of a symmetrical FODO lattice.

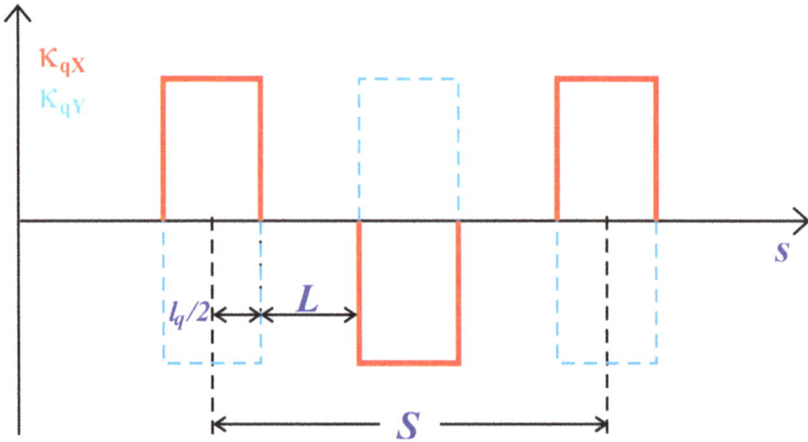

Figure 3.2. Focusing functions of one-and-a-half cells of a *symmetrical* FODO lattice. Note that $\kappa_{qX} = -\kappa_{qY}$; focusing is performed in the horizontal plane and defocusing in the vertical plane. The full-lattice period is 'S.'

In the same vein as equation (3.1) for a solenoid lens, a quadrupole that focuses in the *horizontal or x-plane* is represented by the matrix

$$R_F = \begin{bmatrix} \cos\theta & \dfrac{l_q}{\theta}\sin\theta \\ -\dfrac{\theta}{l_q}\sin\theta & \cos\theta \end{bmatrix}, \tag{3.9}$$

where $\theta = +\sqrt{\kappa_q}\,l_q$, κ_q is positive and given by equation (2.21) for a magnetic quadrupole, and l_q is the effective length of the quadrupole. If the quadrupole focuses in the *vertical or y-plane*, it *defocuses* in the horizontal plane, with a corresponding matrix given by

$$R_D = \begin{bmatrix} \cosh\theta & \dfrac{l_q}{\theta}\sinh\theta \\ \dfrac{\theta}{l_q}\sinh\theta & \cosh\theta \end{bmatrix}. \tag{3.10}$$

Therefore, the matrix corresponding to one symmetrical FODO cell, as shown in figure 3.2, is

$$R_{\text{FODO}} = R_L R_D R_L R_F = \begin{bmatrix} A & B \\ C & D \end{bmatrix}. \tag{3.11}$$

Matrix multiplication leads to

$$
\begin{aligned}
\cos \sigma_0 &= \frac{1}{2} Tr[\boldsymbol{R}_{\text{FODO}}] \\
&= \cos \theta \cosh \theta + \frac{1}{\eta}\theta[\cos \theta \sinh \theta - \sin \theta \cosh \theta] - \frac{1}{2\eta^2}\theta^2 \sin \theta \sinh \theta,
\end{aligned}
\tag{3.12}
$$

with

$$
\eta = \frac{l_q}{L} = \frac{l_q}{(S/2) - l_q},
\tag{3.13}
$$

which defines the *fill factor*. The thin-lens version of equation (3.12) is

$$
\cos \sigma_0 \cong 1 - \frac{1}{2\eta^2}\theta^4 = 1 - \frac{1}{2}\left(L\kappa_q l_q\right)^2.
\tag{3.14}
$$

Since $\cos a = 1 - 2\sin^2(a/2)$ and $f = 1/\kappa_q l_q$, we find

$$
\sin\left(\frac{\sigma_0}{2}\right) = \frac{L}{2f} = \frac{S}{4f},
\tag{3.15}
$$

in terms of the full-lattice period S (figure 3.2). Therefore, $L \leqslant 2f$ for stable motion to occur. We found in chapter 1 that a combination of positive and negative thin lenses with the same strength leads to net focusing if their separation is smaller than one focal length (equation (1.14)). Therefore, the condition for *stable motion* alone is not as restrictive as that for net focusing in a doublet.

As an example, let us consider a *symmetric* periodic FODO lattice with magnetic quadrupoles that have an effective length of $l_q = 0.0516$ m and a full-lattice period of $S = 0.32$ m. If we specify a phase advance per period of $\sigma_0 = 30°$, we obtain $\theta = 0.4341$ from equation (3.12) and $\kappa_q = 70.78$ m^{-2} (see the definitions below equation (3.9)). Note that $\kappa_q = 70.78$ m^{-2} is the value of the focusing (piecewise) constant in the horizontal x-plane; correspondingly, the focusing (piecewise) constant in the vertical y-plane is $-\kappa_q$. Ray propagation in this lattice is illustrated in figure 3.3 for the x and y components of just one ray.

If the strengths of the two quadrupoles in a FODO cell are different, the FODO lattice is *asymmetrical*. The focusing function for such a lattice can be represented as shown in figure 3.4.

In this case, we need to define separate strength variables θ for the two quadrupoles (figure 3.4): $\theta_1 = +\sqrt{\kappa_1}\, l_q$, $\theta_2 = +\sqrt{\kappa_2}\, l_q$. Given these definitions, matrix multiplication, as for equation (3.11), leads to [1]:

$$
\begin{aligned}
\cos(\sigma_{0x}) =\ & \cosh(\theta_2)\left[\cos(\theta_1) - \frac{1}{\eta}\theta_1 \sin(\theta_1)\right] \\
& + \sinh(\theta_2)\sin(\theta_1)\left[\frac{1}{2}\left(\frac{\theta_2}{\theta_1} - \frac{\theta_1}{\theta_2}\right) - \frac{1}{2\eta^2}\theta_1\theta_2 + \frac{1}{\eta}\theta_2 \cot(\theta_1)\right],
\end{aligned}
\tag{3.16}
$$

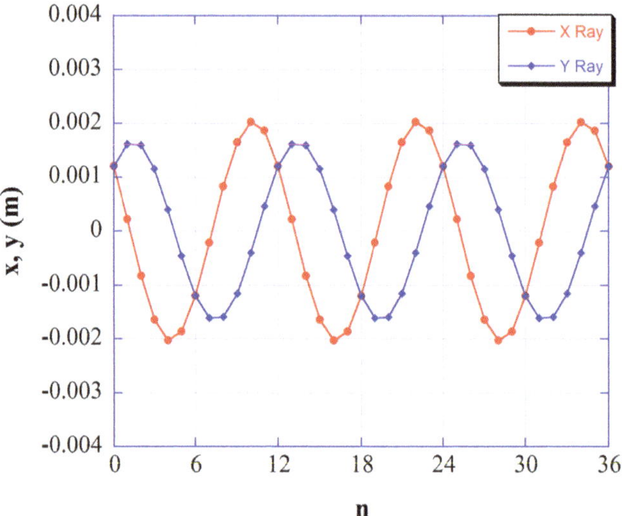

Figure 3.3. Example of ray propagation in a *symmetrical* FODO lattice with 36 periods. Both the x and y components of a ray starting with components $(x_0, x'_0) = (0.0012, 0)$ m and $(y_0, y'_0) = (0.0012, 0)$ m have phase advances of 30° per period.

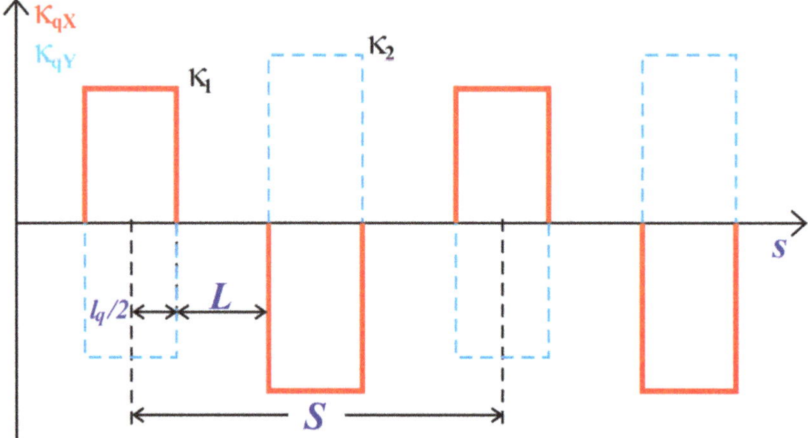

Figure 3.4. Focusing functions of two cells of an *asymmetrical* FODO lattice. Note that $\kappa_1 \neq \kappa_2$, but κ_{qx} still equals $-\kappa_{qy}$ for *each* quadrupole.

$$
\cos(\sigma_{0y}) = \cos(\theta_2)\left[\cosh(\theta_1) + \frac{1}{\eta}\theta_1 \sinh(\theta_1)\right]
$$

$$
+ \sinh(\theta_1)\sin(\theta_2)\left[\frac{1}{2}\left(\frac{\theta_1}{\theta_2} - \frac{\theta_2}{\theta_1}\right) - \frac{1}{2\eta^2}\theta_1\theta_2 - \frac{1}{\eta}\theta_2 \coth(\theta_1)\right]. \tag{3.17}
$$

As an example, let us consider a periodic FODO lattice with the same geometry as before but with different quadrupole strengths such that the phase advances per

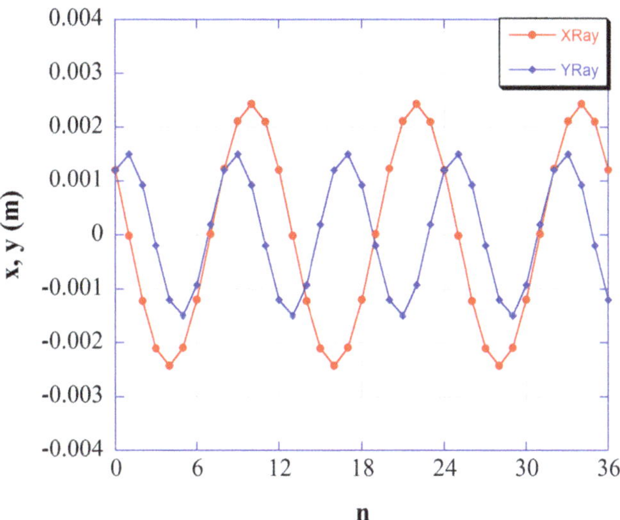

Figure 3.5. Example of ray propagation in an *asymmetrical* FODO lattice with 36 periods. The x-component of a ray starting with components $(x_0, x'_0) = (0.0012, 0)$ m and $(y_0, y'_0) = (0.0012, 0)$ m has a phase advance of $30°$ per period, while the y-component has a phase advance of $45°$ per period.

period in the two transverse planes are $\sigma_{0X} = 30°$, and $\sigma_{0Y} = 45°$. By solving equations (3.16) and (3.17) simultaneously for θ_1, θ_2, we obtain $\theta_1 = 0.4748$ and $\theta_2 = 0.50135$, which correspond to $\kappa_1 = 84.67$ m^{-2} and $\kappa_2 = 94.40$ m^{-2}. Clearly, a stronger second quadrupole in the FODO scheme of figure 3.4 is needed in order to obtain a larger phase advance in the vertical y-plane.

3.3 Lattice and beam functions

The matrix treatment of the previous two sections describing paraxial ray propagation in periodic lattices is equivalent to solving the second-order homogeneous differential equations

$$x''(s) + \kappa_x(s)x = 0,$$
$$y''(s) + \kappa_y(s)y = 0. \tag{3.18}$$

For a solenoid lattice, only one equation would be needed, with the function $\kappa(s)$ representing focusing in the radial direction. Inspired by the simple harmonic oscillator equation, the general solution of the first equation of (3.18) can be written as [2]

$$x(s) = \mathcal{C}w(s)\cos[\psi(s) + \phi], \tag{3.19}$$

where $w(s)$ is the *amplitude function*, $\psi(s)$ is the *phase function*, and $\mathcal{C}^2$ and ϕ are constants that depend on the initial conditions $x(0)$ and $x'(0)$. Differentiation of equation (3.19) and substitution into the first equation of (3.18) yields

$$\gamma x^2 + 2\alpha xx' + \beta x'^2 = \mathcal{C}^2, \tag{3.20}$$

where α, β, and γ define the *Courant–Snyder (C–S) or Twiss parameters*, and $\mathcal{C}^2$ is the *Courant–Snyder invariant*, sometimes also called 'single-particle emittance.'

The Courant–Snyder parameters are functions of s and can be related to the amplitude function and its derivative as follows:

$$\beta = w^2, \quad \alpha = -ww' = -\frac{1}{2}\beta',$$

$$\gamma = \frac{1}{w^2} + w'^2 = \frac{1 + \alpha^2}{\beta}. \tag{3.21}$$

(It will be clear from the context of any discussion whether β and γ refer to the standard relativistic quantities or to the C–S parameters.) Only two of the C–S parameters are independent, because they are connected by the equation

$$\beta\gamma - \alpha^2 = 1, \tag{3.22}$$

which can be easily verified using equations (3.21).

Two additional important equations can be easily obtained from equations (3.19) and (3.18):

$$w''(s) + \kappa_x(s)w(s) - \frac{1}{w(s)^3} = 0, \tag{3.23}$$

and

$$\psi'(s) = \frac{d\psi(s)}{ds} = \frac{1}{\beta(s)}. \tag{3.24}$$

Equation (3.20) represents a tilted ellipse in *trace space* (x, x'), although it is common to call this space 'phase space.' Figure 3.6 shows the ellipse and its relation to the C–S parameters. The ellipse size and shape at a given s are determined by the constant $\mathcal{C}^2$ and the coefficients $\alpha(s)$ and $\beta(s)$. In a periodic structure, the particle coordinates (x, x') trace the ellipse, as illustrated in figure 3.7. We have used the same parameters to calculate the ray propagation shown in figure 3.1. Alternatively, we can construct the trace-space ellipses from equation (3.20) and the C–S parameters. These parameters can be related to the ABCD matrix in equation (3.2) by the following relations:

$$\alpha(s) = \frac{A - D}{2\sin\sigma_0}, \quad \beta(s) = \frac{B}{\sin\sigma_0}, \quad \gamma(s) = -\frac{C}{\sin\sigma_0}. \tag{3.25}$$

(See also the expressions in the discussion before equation (3.7).) The values of the C–S invariant $\mathcal{C}^2$ in equation (3.20) can be determined from the initial conditions. This is shown numerically in one of the examples mentioned in the section on 'Computer Resources' below. The relations in equation (3.25) are a special case of

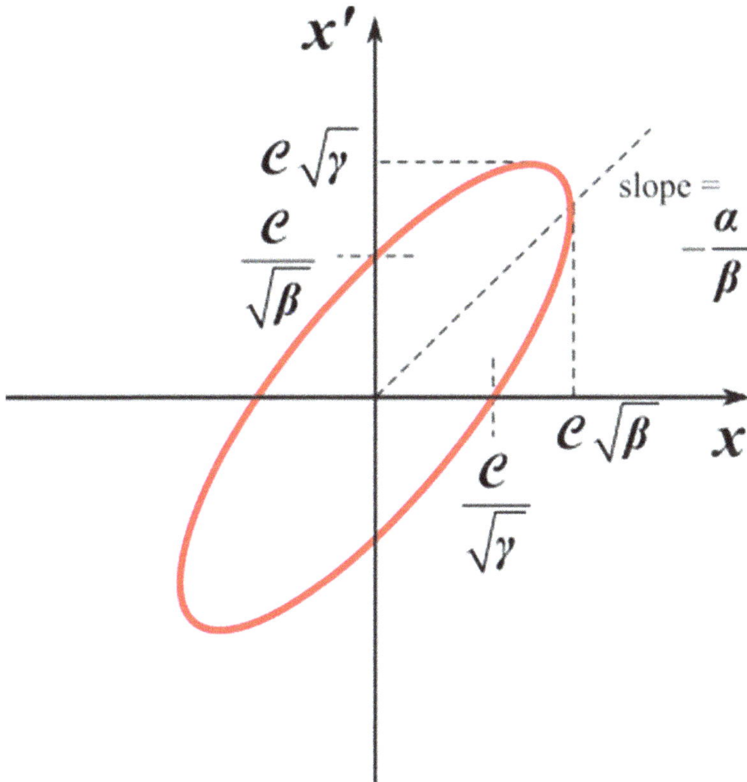

Figure 3.6. Trace-space ellipse (equation (3.20)) in the horizontal plane at a given location '*s*' along the beamline. The size and orientation of the ellipse are related to the C–S parameters (equations (3.21)).

the general matrix for *linear* ray transformation between two planes, 1 and 2, in terms of the corresponding C–S parameters:

$$R_{1\to 2} = \begin{bmatrix} \sqrt{\dfrac{\beta_2}{\beta_1}}(\cos \Delta\psi + \alpha_1 \sin \Delta\psi) & \sqrt{\beta_2\beta_1}\,\sin \Delta\psi \\[2ex] \dfrac{(\alpha_1 - \alpha_2)\cos \Delta\psi - (1 + \alpha_1\alpha_2)\sin \Delta\psi}{\sqrt{\beta_2\beta_1}} & \sqrt{\dfrac{\beta_1}{\beta_2}}(\cos \Delta\psi - \alpha_2 \sin \Delta\psi) \end{bmatrix}, \quad (3.26)$$

where $\Delta\psi$ is the phase advance from plane 1 to plane 2. The derivation of equation (3.26) employs equation (3.19) and its derivative; details can be found in many textbooks (e.g., [3, 4]). For a *periodic lattice*, the matrix in equation (3.26) can be conveniently rewritten as:

$$R_{\text{periodic}} = e^{\mathcal{J}\Delta\psi} = \mathbf{1}\cos \Delta\psi + \mathcal{J}\sin \Delta\psi, \quad (3.27)$$

where $\mathbf{1}$ represents a 2×2 identity matrix and $\mathcal{J}$ represents the matrix $\begin{bmatrix} \alpha & \beta \\ -\gamma & -\alpha \end{bmatrix}$. It follows from equation (3.22) that $\mathcal{J}^2 = -\mathbf{1}$.

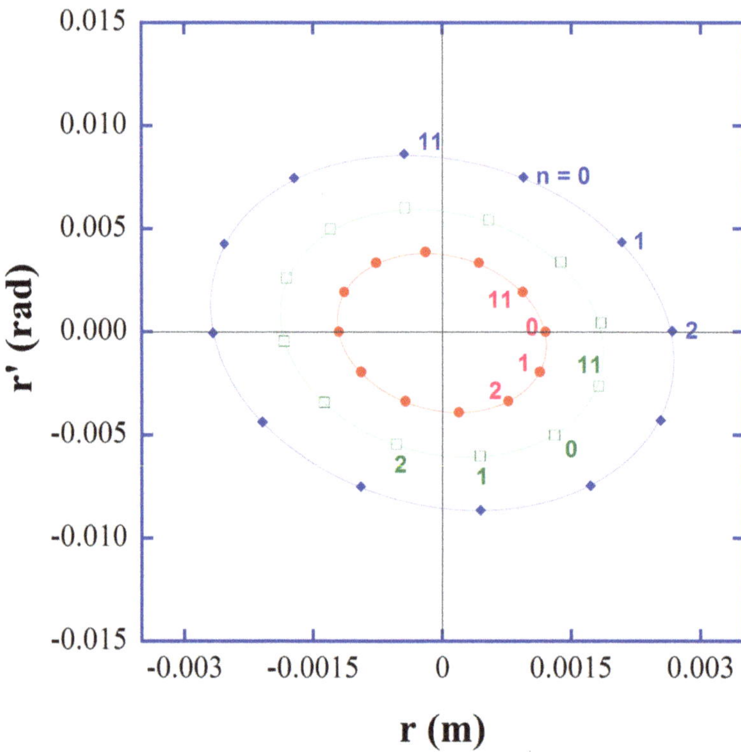

Figure 3.7. Trace-space ellipses corresponding to the rays in figure 3.1 for a periodic solenoid lattice; n is the period number. Note that the points overlap after 12 periods because the phase advance per period is $\sigma_0 = 30°$.

For later reference, we apply equation (3.26) to two cases: the *one-period* and *one-turn* matrices. If we consider the transformation between the midplanes of two horizontally focusing quadrupoles in a symmetrical FODO lattice (see figure 3.2, and figure A.2 in appendix A), we have $\beta_{X1} = \beta_{X2} = \beta_{X\mathrm{max}}$, $\alpha_1 = \alpha_2 = 0$, and $\Delta\psi = \sigma_{0X}$. Similarly, the transformation between the midplanes of two vertically focusing quadrupoles would have $\beta_{Y1} = \beta_{Y2} = \beta_{X\mathrm{min}}$, $\Delta\psi = \sigma_{0Y}$. Assuming that $\sigma_{0X} = \sigma_{0Y} = \sigma_0$, we can find the one-period matrices from equation (3.26):

$$R_X = \begin{bmatrix} \cos \sigma_0 & \beta_{X\,\mathrm{max}} \sin \sigma_0 \\ -\dfrac{\sin \sigma_0}{\beta_{X\,\mathrm{max}}} & \cos \sigma_0 \end{bmatrix}, \quad R_Y = \begin{bmatrix} \cos \sigma_0 & \beta_{X\,\mathrm{min}} \sin \sigma_0 \\ -\dfrac{\sin \sigma_0}{\beta_{X\,\mathrm{min}}} & \cos \sigma_0 \end{bmatrix}, \tag{3.28}$$

which are essentially the same as equation (3.9). After some straightforward algebra, we find that in the *thin-lens* approximation, $\beta_{X\,\mathrm{max},\ X\,\mathrm{min}} \sin \sigma_0 = 2L[1 \pm (L/2f)] = 2L[1 \pm \sin(\sigma_0/2)]$, where we have used equation (3.15) in the last equality. Therefore,

$$\beta_{X\,\mathrm{max}} = \frac{2L[1 + \sin(\sigma_0/2)]}{\sin \sigma_0}, \quad \beta_{X\,\mathrm{min}} = \frac{2L[1 - \sin(\sigma_0/2)]}{\sin \sigma_0}. \tag{3.29}$$

For the *one-turn* matrix, we have, starting and ending at the midplane of a focusing quadrupole, $\beta_{X1} = \beta_{X2} = \beta_{X\text{max}}$, $\alpha_1 = \alpha_2 = 0$, and $\Delta\psi = N\sigma_{0X} = 2\pi\nu_{0X}$, where N is the number of FODO cells in one turn and ν_{0X} is the horizontal 'tune' or number of horizontal betatron oscillations per turn (equation (3.34)). Given these conditions, equation (3.26) becomes

$$\boldsymbol{R}_{\substack{\text{one–turn}\\ \text{hor.}}} = \begin{bmatrix} \cos\left(2\pi\nu_{0x}\right) & \beta\,\sin\left(2\pi\nu_{0x}\right) \\ -\dfrac{\sin\left(2\pi\nu_{0x}\right)}{\beta_{X\,\text{max}}} & \cos\left(2\pi\nu_{0x}\right) \end{bmatrix}. \tag{3.30}$$

So far, we have discussed the propagation of a single particle or ray. In a particle *beam*, different particles have different values of $\mathcal{C}^2$ and phase ϕ in equation (3.19), leading to different ellipse sizes at a location 's.' However, the functions $\alpha(s)$ and $\beta(s)$ depend entirely on the lattice geometry and magnet strengths through the function $\kappa_x(s)$ and for this reason are called *machine* or *lattice functions*.

Equation (3.19) is valid for periodic or non-periodic lattices; only when $\kappa_x(s) = \kappa_x(s + S)$ do we have periodic focusing. Naturally, in this latter case, the C–S parameters are also periodic, indicating that *at a given instant* the trace-space ellipse repeats itself with period S along the reference trajectory. However, a beam injected into a periodic system does not follow the machine ellipse unless the beam is *matched*. When the beam is mismatched, we can still define *beam functions* $\alpha_B(s)$, $\beta_B(s)$, and $\gamma_B(s)$ and a corresponding *beam ellipse*, to be distinguished from the *machine* (or *lattice*) counterparts. The beam functions can be identified with those of the particle with the maximum C–S invariant, but a proper discussion of beam matching normally involves the *beam emittance* and a *beam ellipse* in the manner described in chapter 7.

3.4 Uniform-focusing ('smooth') approximation

Given equation (3.24), we can write an expression for the single-particle *phase advance per period*, introduced in section 3.1 in the matrix treatment of the solenoid lattice, but now made more general in the context of the Courant–Snyder theory:

$$\sigma_0 = \int_{\text{Per.}} \psi'(s)ds = \int_{\text{Per.}} \frac{ds}{\beta(s)}. \tag{3.31}$$

Furthermore, we can substitute an effective *uniform-focusing* for the periodic lattice. The uniform-focusing lattice has a constant focusing function κ_0 and a constant amplitude function w. If we set $w'' = 0$, we use equation (3.23) to find an expression for κ_0 in terms of the constant amplitude function, or the betatron parameter β:

$$\kappa_0 = \frac{1}{w^4} = \frac{1}{\beta^2}. \tag{3.32}$$

We can now relate the phase advance per period of the actual lattice (with period S) to the constant focusing function of the uniform-focusing lattice. According to equation (3.31), the integral over one period of the inverse betatron function can be written as:

$$\sigma_0 = \frac{S}{\bar{\beta}} = S\sqrt{\kappa_0}. \tag{3.33}$$

Finally, we introduce $k_0 \equiv \sqrt{\kappa_0}$, so $\sigma_0 = k_0 S$. An additional related concept, the *betatron tune* or number of betatron oscillations per turn, can be defined for a circular lattice of radius ρ_m,

$$\nu_0 = \frac{1}{2\pi}\oint \frac{ds}{\beta(s)} = \frac{\rho_m}{\bar{\beta}}, \tag{3.34}$$

which can also be written as $k_0 = \nu_0/\rho_m$.

The uniform-focusing or 'smooth' approximation is a very useful construction for theoretical as well as computational studies in accelerator and beam physics. We will return to this model when we consider emittance and space charge in the next chapter.

3.5 Linear dispersion

The single-particle equations of motion (3.18) assume mono-energetic particles, so that the focusing represented by $k(s)$ is defined for a single momentum, p_0. If, however, there is a momentum spread Δp around the design value p_0, the equations are modified. For the horizontal motion (radial in a circular machine), we have the inhomogeneous equation (see, for example, [5] page 46)

$$x''(s) + \kappa_x(s)x = \frac{1}{\rho(s)}\delta, \tag{3.35}$$

where $\delta = \Delta p/p_0$, and $\rho(s)$ is the *local* orbit radius. Here, a major assumption is that $x(s) \ll \rho(s)$, so we can neglect the *weak focusing* term $1/\rho^2$ discussed in section 2.4. The general solution of equation (3.35) is

$$x(s) = x_\beta(s) + D_{0x}(s)\delta, \tag{3.36}$$

where $x_\beta(s)$ is the solution of the *homogeneous* equation (3.18), and $D_0(s)$ defines the *linear dispersion function*. $D_0(s)$ satisfies the equation (see, for example, [5] page 49)

$$D_{0x}''(s) + \left[\frac{1}{\rho^2(s)} + \kappa_x(s)\right]D_{0x}(s) = \frac{1}{\rho(s)}. \tag{3.37}$$

Normally, the weak focusing term $1/\rho^2(s)$ can be neglected, as in equation (3.35). Thus, using the smooth approximation whereby $D_{0x}''(s) \to 0$, we obtain the following for the *average dispersion*:

$$\bar{D}_{0x} = \frac{1}{\kappa_x \rho_m} = \frac{\rho_m}{\nu_0^2}, \tag{3.38}$$

the last equality of which follows the results of the previous section. Therefore, larger circular machines normally need to operate at larger tunes to reduce dispersion effects.

It is straightforward to show, from the definition of magnetic rigidity in equation (2.2), that the momentum error δ is equivalent to a *magnet error*, i.e., $\Delta p/p_0 = \Delta B/B_0$. Therefore, an equation of the same form as equation (3.35) should be solved when studying the effects of magnet errors. We will return to this topic when we discuss betatron resonances (section 7.3) and the closed orbit (section 8.3).

3.6 Momentum compaction, transition gamma, and chromaticity

In general, the change in orbital radius due to momentum error cannot be neglected. Thus, we define the *momentum compaction factor* as the fractional change in equilibrium radius per fractional change in momentum:

$$\alpha_c \equiv \frac{d\rho/\rho_0}{dp/p_0},\tag{3.39}$$

where p_0, ρ_0 define design values. Rewriting equation (2.2) for the magnetic rigidity in the form

$$\rho(p, B) = \frac{p}{qB}, \quad \rho_0 = \frac{p_0}{qB_0},\tag{3.40}$$

and taking logarithmic derivatives on both sides yields $d\rho/\rho_0 = dp/p_0 - dB/B_0$. (Note that we are evaluating the derivatives, i.e., $\partial\rho/\partial p$, $\partial\rho/\partial B$, at the equilibrium values p_0, ρ_0.) Therefore, using the definition of the *field index*, equation (2.26), we find:

$$\frac{d\rho}{\rho_0}(1 - n) = \frac{dp}{p_0}.\tag{3.41}$$

The momentum compaction is then

$$\alpha_c = \frac{1}{1 - n}.\tag{3.42}$$

Furthermore, if C denotes the length of the orbit for a particle of coordinate x, and C_0 is the length of the reference orbit (for which $x = 0$, by definition), then $\alpha_c = (d\rho/\rho_0)\delta^{-1} = (dC/C_0)\delta^{-1}$, and to first order in x we can write (see, for example, [3, 6])

$$\alpha_c = \frac{1}{C_0}\int_0^{C_0}\frac{D_{0x}(s)}{\rho(s)}ds,\tag{3.43}$$

since $dC = C - C_0 = \int_0^{C_0}(x/\rho)ds$, ($\rho(s) =$ the local radius of curvature), and $x(s) = D_{0x}(s)\delta$, from equation (3.36) with $x_\beta = 0$.

Furthermore, because of the change in ρ, there are corresponding changes in the revolution period $T = 2\pi\rho/\beta c$ and the angular frequency $\omega = \beta c/\rho$. Taking logarithmic derivatives in the equation for ω, we get $d\omega/\omega_0 = d\beta/\beta_0 - d\rho/\rho_0$. The second term is $d\rho/\rho_0 = \alpha_c dp/p_0$, according to equation (3.41), while the first term is

$d\beta/\beta_0 = \gamma_0^{-2}dp/p_0$, which follows from the definition of the relativistic mass factor γ, and $d\gamma/\gamma_0 = \beta_0\gamma_0^2 d\beta$ [7]. Finally, the fractional change in period is

$$\frac{dT}{T_0} = -\frac{d\omega}{\omega_0} = \eta_{tr}\frac{dp}{p_0}, \quad \text{with } \eta_{tr} = \alpha_c - \frac{1}{\gamma_0^2}. \tag{3.44}$$

The momentum compaction also defines a *transition gamma*, γ_t

$$\gamma_t = \frac{1}{\sqrt{\alpha_c}}. \tag{3.45}$$

In a straight machine, $\alpha_c = 0$, $\gamma_t \to \infty$, and $\eta_{tr} = -1/\gamma_0^2$: at all energies, particles with higher momentum take less time to traverse a given distance. In circular machines, the same would apply for energies such that $\gamma_0 < \gamma_t$. However, there are situations in which, counterintuitively, particles with higher momentum take *more* time to complete a revolution, i.e., $\eta_{tr} > 0$, or $\gamma_0 > \gamma_t$. This is the regime of 'negative mass' of weak focusing machines such as cyclotrons, or strong focusing circular machines in the regime above the transition energy. For cyclotrons we have $0 < n < 1$, $\nu_r < 1$ (equation (2.30)), while $\nu_r > 1$ for strong focusing machines.

In circular machines, another effect of a momentum error δ is to change the *tunes* in both the radial and vertical planes. Horizontal and vertical *chromaticities*, ξ_X, ξ_Y, are defined to characterize the effect:

$$\Delta\nu_X = \xi_X\delta, \quad \Delta\nu_Y = \xi_Y\delta. \tag{3.46}$$

The focusing of particles with higher momentum is less efficient, leading to reduced tunes. Therefore, we expect the chromaticity produced by linear elements such as dipoles and quadrupoles to be negative.

In a weak focusing machine in which $\partial B_y/\partial\rho$ is constant, the field index satisfies $dn/d\rho = n(1 + n)/\rho$ (see [8]) and we find that

$$\xi_X = -\frac{n(n + 1)}{2(1 - n)^{3/2}}, \quad \xi_Y = -\frac{\sqrt{n}(n + 1)}{2(1 - n)}. \tag{3.47}$$

Note that in Reiser's book [8], the chromaticity is defined in terms of fractional quantities $\Delta\nu_X/\nu_X$, $\Delta\nu_Y/\nu_Y$ instead of just $\Delta\nu_X$, $\Delta\nu_Y$.

A more general definition of chromaticity, referred to as *natural chromaticity* is

$$\xi_X = -\frac{1}{4\pi}\oint\beta_X(s)\kappa_X(s)ds, \quad \xi_Y = \frac{1}{4\pi}\oint\beta_Y(s)\kappa_Y(s)ds, \tag{3.48}$$

where only linear focusing elements are involved. These expressions are easily derived by calculating the first-order differential change in tune when the one-turn matrix (equation (3.30)) is multiplied by a differential-kick error matrix; this latter kick arises from a small fractional momentum error δ. Details can be found in, for example, [9]. (We will encounter a similar expression in chapter 4 when we discuss the effects of linear space charge on the betatron tune; in that case, the kick error is caused by space charge.) Applying equation (3.48) to a FODO lattice of thin lenses and N periods, we obtain [10],

$$\xi_X = -\frac{N}{4\pi}\left(\beta_{X\,\text{max}} - \beta_{X\,\text{min}}\right)\left|\kappa_{qX}\right|l_q = -\frac{2\nu_{0X}}{\sigma_{0X}}\tan\left(\frac{\sigma_{0X}}{2}\right), \tag{3.49}$$

the last equality of which originates from equations (3.29) and (3.15). Thus, the natural chromaticity is ordinarily of the order of the betatron tune but negative. Because large machines operate at large betatron tunes, chromaticity correction is required [10]. As an example, the Spallation Neutron Source (SNS) accumulator ring at Oak Ridge National Laboratory [11] operates at tunes $(\nu_{0X},\,\nu_{0Y}) = (6.23,\,6.20)$; the corresponding natural chromaticities are $(\xi_X,\,\xi_Y) = -(7.9,\,6.9)$. Chromaticity correction is realized using 20 sextupole magnets. We briefly discuss sextupole magnets and chromaticity correction in appendix B.

3.7 Computer resources

Two *Mathcad* files, **Sol_Lattice.xmcd** and **Quad_Lattice.xmcd,** provide the tools used to obtain the ray and trace-space plots in this chapter. In addition, input files for the popular software packages MAD-8 and ELEGANT are included that can be used to calculate lattice functions for an example. Additional examples are given in chapter 8 for linacs and rings.

References

[1] Bernal S, Li H, Kishek R A, Quinn B, Walter M, Reiser M and O'Shea P G 2006 RMS envelope matching of electron beams from 'zero' current to extreme space charge in a fixed lattice of short magnets *Phys. Rev. ST Accel. Beams* **9** 064202

[2] Courant E D and Snyder H S 2000 Theory of the alternating-gradient synchrotron *Ann. Phys.* **281** 360–408 reprint of original 1958 article

[3] Wille K 2000 *The Physics of Particle Accelerators: An Introduction* (New York: Oxford University Press)

[4] Edwards D A and Syphers M J 2004 *An Introduction to the Physics of High Energy Accelerators* (Weinheim: Wiley-VCH)

[5] Bryant P J and Johnsen K 1993 *The Principles of Circular Accelerators and Storage Rings* (Cambridge: Cambridge University Press)

[6] Wolski A 2014 *Beam Dynamics In high Energy Particle Accelerators* (London: Imperial College Press)

[7] Bovet C, Gouiran R, Gumowski I and Reich K H 1970 A selection of formulae and data useful for the design of A.G. synchrotrons, CERN Technical Note DL/70/4, 23 April 1970 https://cds.cern.ch/record/000104153

[8] Reiser M 2008 *Theory and Design of Charged Particle Beams* 2nd edn (Weinheim: Wiley-VCH)

[9] Guiducci S Chromaticity *CAS CERN Accelerator School, CERN* **91-04** 53–66

[10] Montague B W Chromatic effects and their first-order correction *CAS CERN Accelerator School, CERN* **87-03** 75–90

[11] Henderson S *et al* 2014 The spallation neutron source accelerator system design *Nucl. Instrum. Methods Phys. Res. A* **763** 610–73

IOP Publishing

A Practical Introduction to Beam Physics and Particle Accelerators (Third Edition)

Santiago Bernal

Chapter 4

Emittance and space charge

The description of the evolution of beams in accelerators and other devices must be extended beyond the single-particle dynamics concepts of the previous three chapters. The first concept we encounter is the *particle distribution* in phase or trace space. For particle ensembles that satisfy Hamilton's equations of motion, i.e., systems without dissipation (from radiation or collisions) or particle losses, the evolution of any representative region in phase space is similar to the motion of an incompressible fluid. This is the essence of *Liouville's theorem* that we introduce in section 4.1. We then discuss *beam emittance*, which is related to area in phase space and thus invariant if the conditions for Liouville's theorem are satisfied. In practice, however, *root-mean-square (rms) emittance* is used, a beam quality factor that can be shown to be conserved under linear transformations and without acceleration. In section 4.2 we give a simple treatment of the Kapchinskij–Vladimirskij (K–V) and thermal distributions, which are mathematical constructs widely used in beam physics to model real distributions. In the next section, we qualitatively discuss the problems encountered by a classical thermodynamics description of charged-particle beams. In section 4.4, we present the K–V *envelope equations*, which embody the simplest macroscopic beam dynamics of *direct ('incoherent') space charge* (SC) in a uniform-focusing lattice. From the K–V envelope equations, we derive in section 4.5 the expression for the *tune shift* caused by direct SC (incoherent tune shift) for beams of circular and elliptical cross sections. Finally, in section 4.6, we discuss the change in tune caused by image forces (coherent tune shift). In presenting the expressions for SC tune shifts, we use the language of the standard literature as well as the notation of more specialized books such as Reiser's [1]. Most of the illustrations and examples are covered in the computer exercises that we describe briefly at the end of the chapter.

4.1 Liouville's theorem and emittance

We mentioned in chapter 3 that the ellipse represented in equation (3.20) and shown in figure 3.6 is a particle's orbit in (x, x') space, i.e., 'trace' space. More generally, *non-interacting* particles in a beam occupy a six-dimensional phase space formed by three space coordinates and three canonical momenta, or (x,y,z, p_x,p_y,p_z). It can be shown from Hamiltonian mechanics that the *density of particles* or the volume occupied by a number of particles in phase space is invariant. These are statements of Liouville's theorem; its derivation (see, for example, section 3.2 in [1]) corresponds to showing that the evolution of particles in phase space is similar to the motion of an incompressible fluid. If the particles interact, the concept can be extended to a $6N$-dimensional space, where N is the number of particles, but we will retain the use of the simpler six-dimensional space.

Furthermore, if the motions in the three coordinate directions are uncoupled, we can consider the projection of the six-dimensional phase space onto any plane, as was done with the ellipse in figure 3.6 for the (x, x') plane. Therefore, we can state Liouville's theorem for the (x, x') plane in the form

$$\iint dx\,dx' = \frac{1}{p}\iint dx\,dp_x = \text{const.}, \tag{4.1}$$

where $p = \gamma m\beta c$ and similarly for the vertical (y, y') plane (β, γ now stand for the relativistic parameters). An ellipse of the general form $ax^2 + 2bxy + cy^2 = d^2$ encloses an area equal to $\pi d^2/\sqrt{ac - b^2}$; therefore, the area enclosed by the ellipse in figure 3.6 is equal to $A = \pi \mathcal{C}^2$ (not in m^2 but in 'm' units, as x' is dimensionless) after using equation (3.22). If the ellipse represents the trajectory in the trace space of a particle with a maximum amplitude of $\mathcal{C}^2$ in an ensemble of particles, the ratio A/π is identified as the beam emittance:

$$\varepsilon_x = \frac{\pi \mathcal{C}^2}{\pi} = \mathcal{C}^2. \tag{4.2}$$

We adopt this definition, which can be found in many other books (e.g., [1, 2]). An alternative definition of emittance adopted by, for example, [3] includes the factor of π, in which case

$$\varepsilon_x = A = \pi \mathcal{C}^2 \quad \text{(alternative definition)}. \tag{4.3}$$

Its units are frequently shown using the form 'π-mm-mrad,' but there is no fundamental reason for factoring π. Unfortunately, the lack of consensus on the definition of emittance is a source of great confusion.

Since it is not possible in practice to establish the maximum amplitude of particle trajectories, the definition of emittance frequently relies on assuming a *Gaussian distribution* for particle positions and momenta. In the (x, p_x) plane, we have

$$f(x, p_x) = \frac{1}{2\pi\sigma_x\sigma_{p_x}} \exp\left(-\frac{x^2}{2\sigma_x^2} - \frac{p_x^2}{2\sigma_{p_x}^2}\right), \tag{4.4}$$

where σ_x, σ_{p_x} represent *standard deviations*, and we have assumed that the average values of x and p_x are zero, which is the case for distributions around design values. The constant in front of the exponential function in equation (4.4) assures us that $\iint f(x, p_x)dx dp_x = 1$. Furthermore, as can be seen in figure 3.6, the maximum value of x is $\mathcal{C}\sqrt{\beta_x}$. Therefore, if we identify this value with σ_x, we can also write the following for the *emittance*, following the definition in equation (4.2):

$$\varepsilon_x = \frac{\sigma_x^2}{\beta_x}. \tag{4.5}$$

From Liouville's theorem as stated in equation (4.1), emittance decreases according to $1/p$ or $1/\beta\gamma$. Therefore, emittance in Hamiltonian beam transport is conserved only in the absence of acceleration or the emission of radiation (see also chapter 6). However, we can define a conserved quantity, *normalized emittance*, by multiplying by $\beta\gamma$ (β, γ are the relativistic parameters):

$$\varepsilon_{xn} = \beta\gamma\varepsilon_x. \tag{4.6}$$

Without a sharp elliptical boundary, we would like to express the emittance as the product of spreads in both x and x' as if we had an elliptical *upright* boundary: $\tilde{\varepsilon}_x = 2\sigma_x 2\sigma_{x'}$. In general, however, we expect a *statistical definition* of emittance to involve a correlation term of the form xx', because the hypothetical boundary is not upright but rotated. As shown in detail in [4], the *effective*, or *4-rms emittance* is

$$\tilde{\varepsilon}_x \equiv 4\varepsilon_{x\text{rms}} = 4\sqrt{\langle x^2\rangle\langle x'^2\rangle - \langle xx'\rangle}, \tag{4.7}$$

where the angle brackets $\langle\rangle$ denote averages in phase space. For example,

$$\langle x^2\rangle = \sigma_x^2 = \frac{1}{N}\sum_{i=1}^{N}(x_i - \langle x\rangle)^2, \quad \langle xx'\rangle = \frac{1}{N}\sum_{i\neq j}^{N}\sum_{j=1}^{N}(x_i - \langle x\rangle)(x'_j - \langle x'\rangle), \tag{4.8}$$

for N particles. Note that if the correlation term $\langle xx'\rangle = 0$, we regain the expression $\tilde{\varepsilon}_x = 2\sigma_x 2\sigma_{x'}$.

To conclude this section, we present an example of a simple trace-space distribution and its evolution in linear and nonlinear focusing channels. The particles originate from a *point source* located at $s = 0$. Thus, the initial trace space is a vertical line in (x, x') space, i.e., it has no spread in x values but some range in initial trajectory slopes. The particles are emitted with initial *randomly* assigned divergence values x' between -1 and 1 rad. In addition, the particles are focused by a *linear* uniform-focusing channel with a focusing constant $\kappa_0 = 100$ m$^{-2} = k_0^2$. The trace-space line rotates clockwise and completes half a revolution after a distance of $s = \lambda_0/2 = \pi/k_0 = 0.314$ m. Figure 4.1(a) depicts the rotation of the trace-space line formed by 100 particles. Note that the rotation spans an oval, but that the area occupied by the particles as well as the emittance is zero. If focusing is nonlinear and has third-order terms of the form $k_0^4 x^3$ and $k_0^2 x'^2 x$ added to the linear focusing $k_0^2 x$, the trace-space line twists as it rotates, eventually spiraling in and occupying an

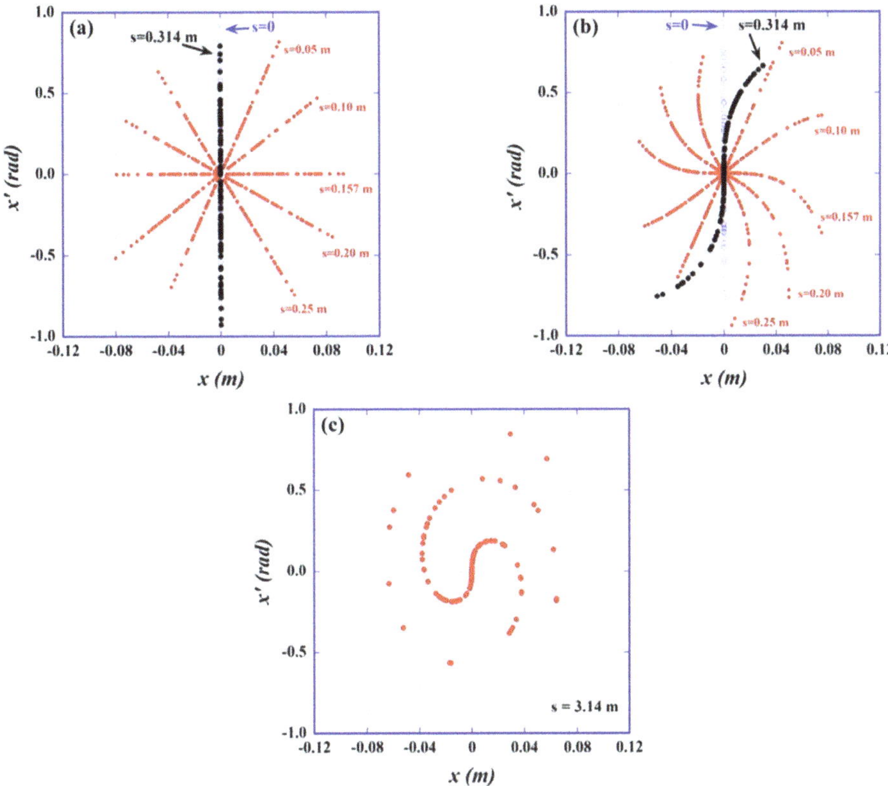

Figure 4.1. Evolution of the trace space of 100 particles originating from a point source at $s = 0$ and moving in (a) a *linear* uniform-focusing channel; (b) a *nonlinear* (cubic) but otherwise uniform-focusing channel; and (c) trace space in a nonlinear channel after five betatron wavelengths. See the text and the Mathcad programs in this book's website.

apparent larger area. Figures 4.1(b) and (c) illustrate the trace-space evolution. The area occupied by the particles as well as the standard emittance is still zero, but the rms emittance now has a non-zero value: the effective emittance, as defined by equation (4.7), is 0.020 m after a distance equal to $\lambda_0/2$ and 0.027 m after a distance equal to $5\lambda_0$. These rather large emittance values stem from the assumed large initial spread in the trajectory slopes; for the linear channel, on the other hand, the effective emittances are 2×10^{-5} m and 3×10^{-4} m after $s = \lambda_0/2$ and $5\lambda_0$, respectively. The calculations and figures can be reproduced using the accompanying *Mathcad* programs. We will consider additional trace-space distributions in the following section.

In many accelerator applications, it is important to have beams with small cross section and beam divergence. Thus, a great deal of accelerator technology is devoted to controlling and minimizing rms emittance. According to Liouville's theorem, emittance is conserved for Hamiltonian *linear or nonlinear processes*. The rms emittance, on the other hand, is conserved only under linear transformations, such as the transformation of coordinates and slopes in equation (1.3) (see also [5]). In most cases, the nonlinearities originate from the focusing magnets, but they may also arise from SC forces. A detailed account of these and other mechanisms of rms emittance growth and theoretical models can be found in chapter 6 of [1]. We cover related additional insights in section 4.3, after we discuss two fundamental particle distributions in beam physics.

4.2 The Kapchinskij–Vladimirskij (K–V) and thermal distributions

Let us consider a continuous, axially symmetric beam with constant particle volume density n_0 (in m^{-3}) within a radius $r = a$. In addition, let us assume that all particles have the same energy and longitudinal velocity $v = \beta c$, with no energy or longitudinal velocity 'errors' or spreads. The beam is kept together by an external *uniform-focusing lattice* characterized by a focusing constant k_0 (in m^{-1}). As we will discover, this arrangement is self-consistent if the beam is 'matched' to the lattice, i.e., if a, n_0, and k_0 are properly chosen. The corresponding particle distribution is an example of a 2D K–V distribution. The K–V distribution is a mathematical construct that can be employed to model the transport of real particle beam distributions with elliptical symmetry [1, 6]. Mathematically we have

$$f_{KV}(p_\perp, r) = \frac{n_0}{2\pi\gamma m}\delta\big[H_\perp(p_\perp, r) - H_0\big],\qquad(4.9)$$

where δ is a Dirac delta function, H_0 is the constant transverse energy, and $H_\perp(p_\perp, r)$ is the transverse Hamiltonian given by

$$H_\perp(p_\perp, r) = \frac{1}{2}\frac{p_\perp^2}{\gamma m} + \frac{1}{2}\gamma m v^2 k_0^2 r^2 + q\phi(r).\qquad(4.10)$$

In equation (4.10), $p_\perp^2 = p_x^2 + p_y^2$, $r^2 = x^2 + y^2$, and $q\phi(r)$ is the self-potential energy due to *linear* SC for particles of charge 'q.' We can imagine a uniform background of the opposite charge to provide a physical basis for the model; magnetic (solenoid) focusing, on the other hand, can be accommodated if the dynamics is described in the Larmor frame (see chapter 3).

From Gauss' law, the self-potential *inside* the beam of radius 'a' can be easily found to be

$$\phi(r) = -\frac{qn_0}{4\varepsilon_0}r^2, \quad 0 \leqslant r \leqslant a.\qquad(4.11)$$

For later reference, we define the *generalized beam perveance* K to be

$$K = \left(\frac{n_0 q^2}{2\varepsilon_0\gamma^3 m}\right)\frac{a^2}{\beta^2 c^2} = \frac{\omega_p^2}{2}\frac{a^2}{\beta^2 c^2},\qquad(4.12)$$

where ω_p is the *plasma frequency*, implicitly given in equation (4.12). Thus, we have

$$q\phi(r) = -\frac{\gamma^3 m v^2 K}{2}\left(\frac{r}{a}\right)^2, \quad 0 \leqslant r \leqslant a.\qquad(4.13)$$

The generalized perveance K can be expressed in terms of the (squared) wave number of plasma oscillations $k_p^2 = \omega_p^2/v^2$, so $K = k_p^2 a^2/2$. Finally, we can rewrite the transverse Hamiltonian in equation (4.10) in the convenient form

$$H_\perp(p_\perp, r) = \frac{1}{2}\frac{p_\perp^2}{\gamma m} + H_0\left(\frac{r}{a}\right)^2, \quad H_0 = \frac{1}{2}\gamma m v^2 a^2\left(k_0^2 - \frac{\gamma^2 k_p^2}{2}\right). \tag{4.14}$$

An alternative form for the K–V particle distribution in equation (4.9) is then

$$f_{KV}(p_\perp, r) = \frac{n_0}{2\pi\gamma m}\delta\left[\frac{1}{2}\frac{p_\perp^2}{\gamma m} - H_0\left(1 - \frac{r^2}{a^2}\right)\right]. \tag{4.15}$$

Using cylindrical coordinates in $(p_\perp, r)$ space and the properties of the δ function, we verify the normalization of the K–V distribution using equation (4.15):

$$2\pi\int f_{KV}\, p_\perp\, dp_\perp = 2\pi\gamma m\int f_{KV}\, d\left(\frac{p_\perp^2}{2\gamma m}\right) = n_0, \quad 0 \leqslant r \leqslant a. \tag{4.16}$$

Furthermore, the *transverse kinetic temperature* in energy units is

$$T_{\perp kin} = \frac{2\pi}{n_0}\int f_{KV}\left(\frac{p_\perp^2}{2\gamma m}\right)p_\perp\, dp_\perp = H_0\left(1 - \frac{r^2}{a^2}\right), \quad 0 \leqslant r \leqslant a. \tag{4.17}$$

The transverse kinetic temperature has a *parabolic profile*, reaching a maximum on the axis, i.e., at $r = 0$, and zero at the beam edge, $r = a$.

As implied at the beginning of this section, the uniform-focusing approximation is employed to simplify the study of the transport of continuous (i.e., 'unbunched') round beams in the presence of significant SC. Under these circumstances, equilibrium transverse distributions are derived that have nonanalytic spatial transverse profiles at given transverse temperatures (see [1], chapter 5). In the limit $T_\perp \to 0$, i.e., the *SC or laminar limit*, the equilibrium distribution is a K–V distribution, which, as we have seen, is uniform in both spatial and velocity coordinates. This limit can also be considered to be the 'zero emittance' limit. However, as we will see below, an equivalent K–V distribution can be defined for *any* combination of emittance and perveance. Furthermore, the limit $T_\perp \to 0$ refers to the limit of a Maxwell–Boltzmann (M–B) distribution; the kinetic temperature, on the other hand, is zero only at the beam edge (equation (4.17)).

In the limit of high temperature (or 'zero current') $T_\perp \to \infty$ *Gaussian (thermal) distributions* are obtained in both spatial and velocity coordinates. The spatial profile has the form of equation (4.4), except that $r = \sqrt{x^2 + y^2}$ is substituted for 'x.' The velocity profile, on the other hand, is Gaussian at all temperatures.

The total, or 100%, and rms emittances in the K–V distribution can be easily shown to be related by ([1], chapter 5)

Table 4.1. Emittance notation.

ε_x	Total emittance in the (x, x') plane
$\varepsilon_{x\text{rms}}$	Root-mean-square (rms) emittance
$\varepsilon_{nx} = \beta\gamma\varepsilon_x$	Normalized emittance (β, γ: relativistic parameters)
$\varepsilon_{nx}^{\text{rms}}$	Normalized rms emittance
$\tilde{\varepsilon}_x = 4\varepsilon_{x\text{rms}}$	Effective emittance ($= \varepsilon_x$ for K–V distribution)

$$\varepsilon_x = 4\varepsilon_{x\text{rms}} \equiv \tilde{\varepsilon}_x, \quad \text{K–V distribution.} \tag{4.18}$$

For a Gaussian distribution, on the other hand, no 100% emittance can be defined because the tails extend to infinity. If the distribution is truncated at n standard deviations σ_x, however, the truncated and rms emittances are related by:

$$\varepsilon_x(n) = n^2\varepsilon_{x\text{rms}}, \quad \text{Gaussian distribution.} \tag{4.19}$$

Following this prescription, the expression in equation (4.5) corresponds to $n = 1$, or 39% of particles: $a^2 \cong \sigma_x^2$. From [3, chapter 3] we find that the fraction of particles F within $n\sigma_x$ is:

$$F = 1 - \exp(-n^2/2) \tag{4.20}$$

For protons, for example, it is customary to use either the 87% ($n = 2.0$) emittance, or the 95% ($n = 2.45$) emittance. We find:

$$\varepsilon_x(87\%) = \frac{4\sigma_x^2}{\beta_x}, \quad \varepsilon_x(95\%) = \frac{6\sigma_x^2}{\beta_x}. \tag{4.21}$$

The beam radius squared corresponding to 95% of the particles in the Gaussian distribution is $a^2 \cong 6\sigma_x^2$.

To recapitulate and for reference, we summarize the emittance notation in table 4.1.

4.3 Thermodynamics of charged-particle beams?

The K–V distribution is an example of a *microcanonical* distribution in statistical mechanics, i.e., a particle distribution constructed for an isolated system at constant energy. However, other related concepts from statistical mechanics, such as *partition function* and *entropy*, are rarely discussed in connection with charged-particle beams because of problems in identifying dynamical 'states' for interacting particles in a beam, even in the simple context of the K–V distribution. More generally, there are conceptual difficulties associated with the use of statistical thermodynamics in beam physics.

In a mathematical context, it is the lack of *extensitivity* of the total energy and entropy in charged-particle beams (and also gravitational systems) which in principle disallows a standard thermodynamics treatment (see [7]). If $\Omega(E)$ is proportional to the volume in phase space at a (constant) total energy E, the microcanonical entropy is $S(E) = k_\text{B}\ln\Omega(E)$, where $\Omega(E)$ is also interpreted as a thermodynamic probability. The entropy is extensive, i.e., additive, only if the system can be subdivided into n subsystems such that $\Omega = \Omega_1\Omega_2...\Omega_n$, i.e., if the subsystems are statistically independent. However, a charged-particle beam cannot be arbitrarily partitioned into statistically independent subsystems in the same way that an ideal gas can. The latter condition is required to guarantee that all microscopic states compatible with a constant total energy are equally accessible, i.e., they have the same probabilities.

The transverse rms emittance, $\tilde{\varepsilon}_x$, on the other hand, has been related to temperature [1] and also to entropy [8]. The first case is justified by the use of $\tilde{\varepsilon}_x$ as a measure of the beam divergence which results from transverse kinetic ('thermal') energy, while the connection to entropy arises from the identification of $\tilde{\varepsilon}_x$ with the phase-space area and the number of dynamical 'states' associated with that area. However, in classical equilibrium thermodynamics, temperature is an intensive variable, while entropy is an extensive variable. Extensive variables scale linearly with the number of particles in the system, while intensive variables remain constant.

Another concept widely used to describe charged-particle beams is the *Boltzmann factor* $\exp(-H_\perp/k_\mathrm{B}T_\perp)$. Here, $H_\perp$ is the transverse Hamiltonian of equation (4.14) and $T_\perp$ the *transverse temperature*, which is normally much larger than the *longitudinal temperature* (see [1], chapter 6). The *canonical ensemble* is implicit in this picture, but no discernible *heat bath*, which is a key component of the ensemble, is present. Therefore, $T_\perp$ cannot be a thermodynamic temperature, but simply a *kinetic parameter* characterizing the spread of energies in a model distribution. More importantly, a beam may evolve towards an equilibrium described by an effective M–B distribution, but such equilibrium cannot be rigorously characterized as *thermodynamic equilibrium*.

Nevertheless, it is possible to formulate a phenomenological theory of emittance growth which, on the surface, resembles a thermodynamics framework. The theory, which was developed by Wangler, Reiser, O'Shea, and others [1, 9], can be summarized by the following equation for the rate of change of emittance in an unbunched beam:

$$\frac{d\varepsilon_{xn}^{\mathrm{rms}}}{dz} = -\frac{\gamma\langle x^2\rangle}{mc^2 N\varepsilon_{xn}^{\mathrm{rms}}}\frac{d\hat{F}}{dz}, \quad \hat{F} = \hat{U} - T_{\mathrm{eff}}S_n, \tag{4.22}$$

where $\varepsilon_{xn}^{\mathrm{rms}}$ is the normalized, transverse rms emittance, $\langle x^2\rangle$ is the squared rms transverse dimension at z (equation (4.8)), N is the number of particles per unit length, $\hat{F}$ is a *generalized free-energy function*, reminiscent of the Helmholtz free energy of classical thermodynamics, S_n is the *normalized entropy* ($S/k_\mathrm{B}N$), and T_{eff} is an effective transverse temperature for the beam given by

$$T_{\mathrm{eff}} = \frac{mc^2(\varepsilon_{xn}^{\mathrm{rms}})^2}{k_\mathrm{B}\gamma\langle x^2\rangle}. \tag{4.23}$$

The term $\hat{U} = (U + W) - (U_0 + W_0)$ contains contributions from self-field (U) and transverse kinetic energy (W) changes relative to the effective linear models (K–V beam distribution for the transverse plane). The entropy term in the original formulation [9] is the entropy of the average slice in the bunch; for an unbunched beam, this entropy characterizes the entire beam.

The original phenomenological model of Wangler *et al* [10] is applicable to SC-dominated beam transport (see the next section). Therefore, emittance growth is driven by SC forces and the beam evolves to be one whose distribution is close to a K–V distribution (for which $T_{\text{eff}} = 0$). In this form, the theory has been verified by experiments at the University of Maryland and Los Alamos National Laboratory ([1], chapters 6 and 7). The more general model embedded in equation (4.22) with the additional entropy term is, in principle, applicable to emittance-dominated beam transport as well (for which $T_{\text{eff}}S_n \gg \hat{U}$), but has not been confirmed, either by experiments or computer simulations. The fundamental issue of the conditions for reversibility/irreversibility of emittance growth, which is addressed in [9], remains an open question.

4.4 The K–V envelope equations and space-charge (SC) intensity parameters

Using the *linear* transverse SC forces arising from a K–V distribution, we can derive an equation governing the evolution of the rms beam radius in a linear uniform-focusing channel. From $x''(s) = -k^2 x(s)$ and $X(s) \equiv \sqrt{\langle x^2 \rangle}$, where the 'spring constant' k includes *both* external focusing and internal ('direct') SC, we find, after straightforward differentiation of $X(s)$ and the statistical definition of emittance in equation (4.7) (see [11] for details),

$$X''(s) + k^2 X(s) - \frac{\varepsilon_{x\text{rms}}^2}{X^3(s)} = 0. \tag{4.24}$$

The unnormalized rms emittance $\varepsilon_{x\text{rms}}$ is related to the effective emittance by $\varepsilon_{x\text{rms}} = \tilde{\varepsilon}_x/4$ (equation (4.7)). Equation (4.24) can be derived from equation (3.23) for the amplitude function $w(s)$ of the Courant–Snyder theory discussed in section 3.3. We can write $X(s) = \sqrt{\varepsilon_{x\text{rms}}}\, w(s) = w_{\max}(s)$ to obtain an equation for the maximum amplitude function $w_{\max}(s)$ from equation (3.23). We also assume uniform focusing and perform the substitution $k_0^2 \to k^2$ to include SC defocusing. Therefore, using the *effective beam radius* $R(s) \equiv 2X(s)$ and the *effective emittance* $\tilde{\varepsilon}_x$, equation (4.24) can be cast in the alternative form

$$R''(s) + k^2 R(s) - \frac{\tilde{\varepsilon}_x^2}{R^3(s)} = 0, \tag{4.25}$$

which is the basis for the *beam envelope* calculations discussed in sections 7.1 and 7.2. Note that the equations for either $X(s)$ or $R(s)$, unlike that for $w(s)$, involve the *statistical* quantity $\varepsilon_{x\text{rms}}$, which characterizes a beam distribution.

As mentioned above, 'k' includes both (linear) external focusing and SC defocusing. Explicitly,

$$k^2(s) = k_0^2 - \frac{K}{R^2(s)} \to k_0^2 - \frac{K}{a^2}, \text{ (smooth approximation)}, \tag{4.26}$$

where K is the generalized beam perveance or 'SC parameter' defined in equation (4.12), and k_0 is the focusing constant. The dependence of the linear SC force on the beam radius is discussed in detail by Reiser (see section 5.3.4 in [1]). It turns out that the average SC force is independent of the details of the particle distribution as long as the distribution has elliptical symmetry; when this is the case, an *equivalent K–V distribution* can be defined [1].

Furthermore, $R''(s) = 0$ in the smooth approximation (equation (4.25)), and we have $R = a = \sqrt{\tilde{\varepsilon}_x/k}$ for the 'average' beam radius of the *rms-envelope matched* equivalent K–V beam. The connection between the parameters of a periodic focusing lattice and those of the smooth approximation is made, as in equation (3.33), through $k = \sigma/S$, and $k_0 = \sigma_0/S$, where σ, σ_0 are the phase advances per period S, *with and without SC*, respectively. Note that we now have a new phase advance that includes the effect of linear SC, but that we are *not* explicitly using a betatron function that includes SC.

In figure 4.2 we present an example of calculations of the beam envelope and single-particle trajectories in a periodic solenoid lattice and in the equivalent uniform-focusing lattice. The beam perveance chosen is $K = 0.0015$ (equation (4.12)), while the effective emittances are $\tilde{\varepsilon}_x = \tilde{\varepsilon}_y = 60$ μm.

The lattice period is $S = 0.32$ m, the solenoid effective length is 0.065 m, and the peak focusing constant is $k_0^2 = 100$ m^{-2}. We solve the K–V envelope equation (4.25) with a lattice of hard-edge solenoids and also with uniform focusing with an average focusing constant of $k_{\text{ave}}^2 = 20.31$ m^{-2}. By construction, the smooth-focusing model does not yield exactly the same phase advance and tune depression (see equation (4.27) below) as the piecewise-focusing model (figure 4.2). Details can be found in a *Mathcad* program briefly described at the end of the chapter.

Using equations (4.25) and (4.26), a number of dimensionless *SC intensity parameters* can be defined. The first one is called the *tune depression η*,

$$\eta = \frac{k}{k_0} = \frac{\sigma}{\sigma_0}, \tag{4.27}$$

which is applicable to periodic linear or circular machines, although the word 'tune' is directly relevant to circular machines only. The tune depression, however, is equal to one for 'zero' current ($K = 0$ in equation (4.26)) and to zero for 'zero' emittance.

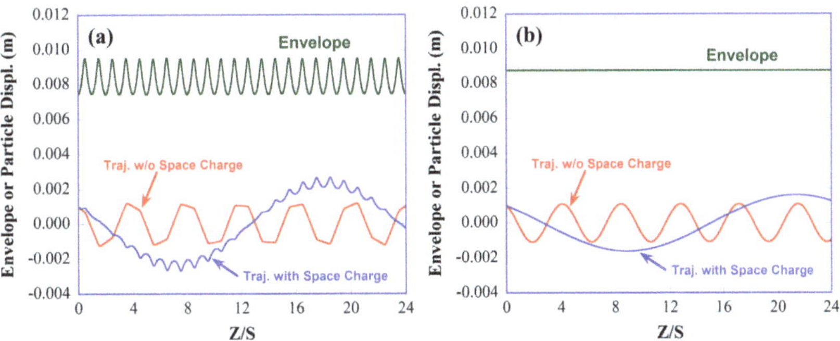

Figure 4.2. (a) Matched beam envelope and single-particle trajectories in a *periodic* solenoid lattice. (b) Matched beam envelope and single-particle trajectories in the equivalent uniform-focusing lattice. The tune depressions (equation (4.27)) are (a) 0.193 and (b) 0.175.

In the latter case we approach the *SC limit* $k \to 0$, whereby external focusing exactly balances direct SC (equation (4.26)), which also implies $\lambda = 2\pi/k \to \infty$, i.e., perfectly laminar flow. Therefore, it is perhaps better to change the definition of an SC intensity parameter to

$$\chi = 1 - \eta^2 = \frac{K}{k_0^2 a^2},\qquad(4.28)$$

as was independently done by Reiser and Davidson (see [1, 6]), so that the range of χ is from zero (in the limit of zero current) to one (in the SC limit). Note that χ is, according to equations (4.25) and (4.26), equal to the ratio of the SC force K/a to the external focusing force $k_0^2 a$ at the effective beam edge $R = a$. Another SC intensity parameter, related to χ and also used by Reiser [1] and Davidson and Qin [6], can be used to more transparently express the role of external focusing and the beam quantities:

$$u = \frac{K}{2k_0\tilde{\varepsilon}_x}.\qquad(4.29)$$

In contrast to χ, u ranges from zero at zero current to infinity at zero emittance. It is straightforward to show the connection between u and χ:

$$\chi = \frac{2}{1 + \sqrt{1 + u^{-2}}}.\qquad(4.30)$$

To complement the equations above, it is interesting to note that the ratio of effective beam radii in the limits of zero emittance (a_{B}) and zero current (a_0) is given by

$$\frac{a_{\mathrm{B}}}{a_0} = \frac{\sqrt{K/k_0^2}}{\sqrt{\tilde{\varepsilon}_x/k_0}} = (2u)^{1/2},\qquad(4.31)$$

as can be easily shown from the envelope equation (4.25). For any combination of SC and emittance, however, the effective 'average' beam radius 'a' is better approximated by just adding a_{B} and a_0 in quadrature:

$$a \cong \sqrt{a_{\mathrm{B}}^2 + a_0^2} = a_0\sqrt{1 + 2u},\qquad(4.32)$$

with the provision that, as an approximation, $u \leqslant 5$ ($\chi = 0.99$ for $u = 5$). The expression in equation (4.32) overestimates the beam radius 'a' by a few percent relative to exact envelope calculations (see section 7.1). The *exact* smooth-approximation results for 'a,' in terms of χ or u, are:

$$a = a_0(1 - \chi)^{-1/4},\quad 0 < \chi < 1,$$
$$a = a_0\sqrt{u + \sqrt{1 + u^2}},\quad u > 0,\qquad(4.33)$$

as can easily be derived from $a = \sqrt{\tilde{\varepsilon}_x/k}$, $a_0 = \sqrt{\tilde{\varepsilon}_x/k_0}$, and the definitions of χ and u in equations (4.28) and (4.29). In general, equation (4.33) underestimates 'a' relative to exact envelope calculations.

We emphasize that the SC intensity parameters, u and χ, are applicable only to *rms-envelope matched* beams in *periodic* lattices. Both parameters are needed to understand the role of SC and how to reduce its effect: for constant current and emittance, we need stronger focusing to make the beam smaller (equation (4.29)), which increases the SC force but also reduces the ratio of the SC force to the external focusing force (equation (4.28)); see also [12]. In the ideal SC limit, this latter ratio is maximized and equal to $\chi = 1$ (or $\eta = 0$), but we consider $\chi > 0.5$ (or $\eta < 0.71$, $u > \sqrt{2}/4 \cong 0.35$) to be the regime of *SC-dominated* beam transport, because in this regime *plasma oscillations* dominate over zero-current betatron motion. To see this in detail, let us use the *plasma wave number* $k_p = 2\pi/\lambda_p = \omega_p/\beta c$, where ω_p is the plasma frequency implicitly defined in equation (4.12). The beam perveance, defined in the same equation (4.12), is then $K = a^2 k_p^2/2$, and we can introduce the *plasma tune depression*,

$$\eta_p \equiv \frac{k_p}{k_0} = \sqrt{2\chi}, \qquad (4.34)$$

in terms of the SC intensity parameter, equation (4.28). For $\chi > 0.5$, we have $\eta_p > 1$, indicating that the plasma wavelength is shorter than the betatron wavelength, $\lambda_p < \lambda_0$. The term 'SC-dominated *beam*' abounds in the literature, but it is more appropriate to speak of 'SC-dominated *beam transport*,' as a beam with given current and emittance can be either emittance- or SC-dominated, depending on external focusing.

In a general *transfer line* (i.e., a nonperiodic section), or for a mismatched beam, a different SC intensity parameter should be used. A natural choice is the ratio of SC to emittance terms in the envelope equation. Using equations (4.25) and (4.26), we can define the parameter S_p:

$$S_p \equiv \frac{KR^2(s)}{\tilde{\varepsilon}_x^2}. \qquad (4.35)$$

Given equations (4.26) and (4.28), the last equation can be written as $\chi/(1 - \chi)$ for an rms-matched beam ('average' radius $R = a$) in a periodic lattice. Collective phenomena dominate over single-particle motion if $S_p \gg 1$, which is equivalent to $a \gg \lambda_D$, where λ_D is the *Debye length* of the beam defined by the relation

$$\lambda_D \omega_p = v_{th} = \sqrt{\frac{k_B T}{m}}, \qquad (4.36)$$

where v_{th} is the transverse rms thermal speed of the charged particles. This speed and the effective beam radius a can be related to the *rms* x-emittance by $\varepsilon_{xrms} = XX' = (a/2)(v_{th}/\beta c)$.

4.5 Incoherent space-charge (SC) betatron tune shift

The tune, i.e., the number of betatron oscillations in one revolution in a circular lattice, changes under the action of direct SC and also from image forces that act when the beam is off-center in the vacuum pipe. The first effect defines the *incoherent tune shift*, which can easily be calculated using the K–V envelope equation. If ν_{0x} is the bare tune (horizontal plane), defined in equation (3.34), and ν_x the *shifted tune* from direct SC, we can write

$$\nu_{0x} - \nu_x = \nu_{0x}\left(1 - \sqrt{1 - \chi}\right), \tag{4.37}$$

using equations (4.27) and (4.28), and $\nu/\nu_0 = \sigma/\sigma_0$. Furthermore, if we assume *weak* SC, i.e., small χ, we have $a \cong a_0 = \sqrt{\tilde{\varepsilon}_x/k_0}$, and with $k_0 = \nu_0/\rho_m$ (see equation (3.34)) we can write

$$\nu_{0x} - \nu_x \cong \frac{\nu_{0x}\chi}{2} = \frac{K\rho_m}{2\tilde{\varepsilon}_x}, \tag{4.38}$$

where 'ρ_m' above is the *machine radius*. Two more commonly quoted equations for the incoherent SC tune shift $\Delta\nu_x \equiv \nu_{0x} - \nu_x$ for *weak* SC are

$$\Delta\nu_x \cong \frac{I\rho_m}{I_0\beta^3\gamma^3\tilde{\varepsilon}_x} = \frac{Nr_e}{2\pi\beta^2\gamma^3\tilde{\varepsilon}_x}, \quad \text{(K–V Distribution)}, \tag{4.39}$$

where we have used the expression for the beam perveance K given in equation (4.12). For the second equality we employ the definition of the classical electron radius $r_e = ec/I_0$, $I_0 = 4\pi\varepsilon_0 mc^3/e$, and $I = Ne\beta c/2\pi\rho_m$ in terms of the total number of particles N of a continuous beam. Other authors use $N/2\pi\rho_m$ for the *number of particles per unit length*. Note that for a given energy and small direct SC the incoherent tune shift is proportional to the beam-current/emittance ratio and is independent of the bare tune ν_{0x}. The scaling of $\Delta\nu_x$ according to $1/\gamma^3$ is, however, the most dramatic, explaining why direct SC effects are negligible in electron machines such as light sources and insignificant in most proton synchrotrons above about 10 GeV [13].

In general, the incoherent tune shift depends on ν_{0x} and becomes comparable to it for very strong SC. Therefore, another sensible measure of SC strength is the ratio of the tune shift to the bare tune:

$$\frac{\Delta\nu_x}{\nu_{0x}} = 1 - \sqrt{1 - \chi} \cong \frac{\chi}{2}, \tag{4.40}$$

which is obtained from equations (4.37) and (4.38). Note that the incoherent tune shift alone would not be a good measure of incoherent SC effects for overall beam transport; by contrast, the relative change in tune can be related to χ or u. For example, an incoherent betatron tune shift of 0.4 at an operating horizontal tune of 6.7, as in the Fermilab booster, yields a relative tune shift of about 6%. While this is significant for resonance crossing (see chapter 7), it does not correspond to very strong transverse SC effects, i.e., beam transport is still dominated by emittance.

Another—more common—approach used to derive the incoherent SC tune shift starts with the Courant–Snyder matrix treatment of a circular lattice and considers the SC effect as a distributed defocusing 'error.' The following tune-shift formula is obtained from a *linear approximation* of the strength of the defocusing error Δk_0^2 (see the discussion related to equation (3.48) and also [3] or [14] for details):

$$\Delta \nu_x = -\frac{1}{4\pi} \oint \beta_x(s) \Delta k_0^2(s) ds, \tag{4.41}$$

where the zero-current beta function is evaluated at the location of the error. Using equation (4.26), we have $\Delta k_0^2(s) = -K/R^2(s)$ (note the sign); we can also define

$$\left\langle \frac{\beta_x(s)}{R^2(s)} \right\rangle = \frac{1}{2\pi\rho_m} \int_0^{2\pi\rho_m} \frac{\beta_x(s)}{R^2(s)} ds. \tag{4.42}$$

We can now identify the average on the right as just $1/\tilde{\varepsilon}_{x\mathrm{rms}}$, where $\beta_x(s)$ is the *zero-current* beta function and $\langle R^2(s) \rangle \cong a_0^2$. Therefore, we recover the result in equation (4.38) for the incoherent tune shift (for weak SC and a uniform particle distribution) by combining equations (4.41) and (4.42).

If the particle distribution is *Gaussian*, as in equation (4.4), we can consider its projection in configuration space by integrating over p_x (and also over p_y after defining $f(y,p_y)$ for the vertical plane). We can then obtain a charge density of the form

$$n_q(r) = \frac{I}{2\pi\beta c\sigma^2} \exp(-r^2/2\sigma^2), \tag{4.43}$$

with $r = \sqrt{x^2 + y^2}$. The defocusing caused by this type of distribution is *nonlinear*, but linearizing the SC force for small 'r,' i.e., by assuming $r \ll \sigma$, leads to a *small-amplitude tune spread* [13, 15] equal to *twice* the size of the tune shift for a uniform distribution that has the same current and rms emittance:

$$\Delta \nu_x \cong \frac{2I\rho_m}{I_0\beta^3\gamma^3\varepsilon_{\mathrm{rms}}} = \frac{2Nr_e}{2\pi\beta^2\gamma^3(4\varepsilon_{\mathrm{rms}})}, \quad \text{(Gaussian distribution)}, \tag{4.44}$$

Note that we now have a *tune spread* instead of a tune shift, because particles moving at different amplitudes have different tunes.

So far, we have assumed that the beam has a circular cross section, but in actual alternating-gradient (AG) focusing the bare tunes in the two transverse directions may have different values, and the beam may have correspondingly different average horizontal and vertical dimensions. Therefore, if the beam has a uniform density but average dimensions 'a' and 'b' in the horizontal and vertical directions, respectively, the horizontal incoherent SC tune shift depends on the parameters for both the horizontal and vertical planes. The *small-amplitude* incoherent tune shift in the vertical direction can be derived along the lines of equations (4.41) and (4.42) (see [13] or [16]):

$$\Delta \nu_y \cong \frac{Nr_e}{\pi\beta^2\gamma^3\tilde{\varepsilon}_y}\left(1 + \sqrt{\overline{\beta_x}\tilde{\varepsilon}_x/\overline{\beta_y}\tilde{\varepsilon}_y}\right)^{-1}. \tag{4.45}$$

The result for $\Delta \nu_x$ (horizontal direction) is obtained by exchanging the x and y subscripts above.

We can calculate incoherent SC tune shifts for elliptical beams and *arbitrary* SC using the K–V envelope equations. In equations (4.25) and (4.26), we presented the K–V envelope equation for an axisymmetric beam in a uniform solenoidal focusing channel. We generalize the equation to piecewise AG focusing (section 3.2) and write (see page 325 in [1])

$$
X''(s) + k_{0x}^2(s)X(s) - \frac{2K}{[X(s) + Y(s)]} - \frac{\tilde{\varepsilon}_x^2}{X^3(s)} = 0,
$$

$$
Y''(s) + k_{0y}^2(s)Y(s) - \frac{2K}{[X(s) + Y(s)]} - \frac{\tilde{\varepsilon}_y^2}{Y^3(s)} = 0,
$$

(4.46)

where we have allowed for the possibility of different focusing functions as well as different effective emittances in the two transverse Cartesian planes. (The equations are often called 'K–V equations' although the K–V distribution is not a self-consistent distribution in an AG focusing lattice.) The corresponding smooth-approximation, *coupled* algebraic equations are

$$
k_{0x}^2 a - \frac{2K}{a + b} - \frac{\tilde{\varepsilon}_x^2}{a^3} = 0,
$$

$$
k_{0y}^2 b - \frac{2K}{a + b} - \frac{\tilde{\varepsilon}_y^2}{b^3} = 0,
$$

(4.47)

with the understanding that all quantities are constants. Thus, following a procedure similar to that used for beams with circular cross sections, we define *tune depressions* $\eta_{x,y}$ in the two transverse planes using the equations

$$
\eta_x \equiv \frac{k_x}{k_{0x}} = \frac{\nu_x}{\nu_{0x}}, \qquad \eta_y \equiv \frac{k_y}{k_{0y}} = \frac{\nu_y}{\nu_{0y}},
$$

(4.48)

and write the following for the tune shifts (see equation (4.37)):

$$
\nu_{0x} - \nu_x = \nu_{0x}\left(1 - \sqrt{1 - \chi_x}\right),
$$

$$
\nu_{0y} - \nu_y = \nu_{0y}\left(1 - \sqrt{1 - \chi_y}\right),
$$

(4.49)

where the intensity parameters $\chi_{x,\,y}$ are given by

$$
\chi_x = \frac{2K}{k_{0x}^2(a + b)a}, \qquad \chi_y = \frac{2K}{k_{0y}^2(a + b)b}.
$$

(4.50)

For weak SC, equation (4.49) and the first equality in the last equation lead to

$$
\Delta\nu_x \cong \frac{\nu_{0x}\chi_x}{2} = \frac{K\rho_m^2}{\nu_{0x}[1 + (b/a)]a^2},
$$

(4.51)

which reduces to equation (4.38) for $a = b$, $\tilde{\varepsilon}_x = \tilde{\varepsilon}_y$. Furthermore, given $b/a \cong b_0/a_0 = \sqrt{\tilde{\varepsilon}_y\overline{\beta}_x/\tilde{\varepsilon}_x\overline{\beta}_y}$, we find

$$\Delta\nu_x \cong \frac{K\rho_m}{\tilde{\varepsilon}_x}\left(1 + \sqrt{\overline{\beta}_y\tilde{\varepsilon}_y/\overline{\beta}_x\tilde{\varepsilon}_x}\right)^{-1}, \tag{4.52}$$

which agrees with the x version of equation (4.45). Note that for arbitrary SC, equation (4.47) must first be numerically solved for 'a' and 'b,' then the intensity parameters are calculated from equation (4.50), and finally the incoherent tune shifts result from equation (4.49). Alternatively, we can solve the K–V envelope equation (4.46) for the piecewise focusing functions and then obtain the phase advances (per period) with and without SC. One of the numerical examples described at the end of the chapter deals with incoherent SC tune shift calculations employing both the algebraic (smooth approximation) and envelope approaches. See also [17].

4.6 Coherent tune shift and Laslett coefficients

Beams require a good vacuum to be transported over long distances without losses or deleterious effects. In addition, bending magnets that use ferromagnetic materials are common in circular machines. Thus, the betatron tune can be significantly affected because particles that are off-axis in the pipe are subject to *image forces.* These forces depend not only on the particle's energy but also on the geometry and magnetic properties of the boundaries. The problem of finding these forces for different beam and pipe geometries was solved by L J Laslett in the 1960s. For a round beam, for example, the formula for the *incoherent* SC tune shift, equation (4.39), must be supplemented with two terms involving the pipe height 'h,' the half-gap 'g' of the bending magnets, and the Laslett coefficients ε_1 and ε_2 for *incoherent* tune shift:

$$\Delta\nu_{x,\,\text{inc.}} \equiv \nu_{0x} - \nu_{x,\,\text{inc.}} \cong \frac{Nr_e}{\pi\beta^2\gamma\tilde{\varepsilon}_x}\left[\varepsilon_1\left(\frac{1}{\gamma^2} + \frac{a_0^2}{h^2}\right) + \varepsilon_2\beta^2\frac{a_0^2}{g^2}\right], \tag{4.53}$$

where $\varepsilon_1 = \frac{1}{2} = \varepsilon_2$ for a circular pipe. Note that we recover equation (4.39) if a_0^2/h^2, $a_0^2/g^2 \ll 1$ (a_0 is the zero-current beam radius). Note also that equation (4.53) is valid for *weak* SC, i.e., for tune shifts that are small compared to the operating (bare) tune; this is the case for almost all accelerators in existence. When beam transport is dominated by transverse SC effects, on the other hand, equation (4.37) must be used to calculate the incoherent tune shift; we give examples for both weak and strong SC in chapters 8 and 9.

The effect on the beam centroid, i.e., on the coherent motion, leads to a *coherent* SC tune shift given, for *penetrating fields*, by the so-called 'integer formula' (see [13], [18], or [19]):

$$\Delta\nu_{x,\,\text{coh.}} \equiv \nu_{0x} - \nu_{x,\,\text{coh}} \cong \frac{Nr_e}{\pi\beta^2\gamma\tilde{\varepsilon}_x}\left[\xi_1\frac{a_0^2}{h^2} + \xi_2\beta^2\frac{a_0^2}{g^2}\right], \quad \text{for penetrating fields,} \tag{4.54}$$

where ξ_1 and ξ_2 are the Laslett coefficients for *coherent* tune shift. For a round pipe, $\xi_1 = \frac{1}{2}$, $\xi_2 = \pi^2/16 = 0.617$. For *non-penetrating fields*, the 'half-integer formula' applies:

$$\Delta\nu_{x,\,\text{coh.}} \cong \frac{Nr_e}{\pi\beta^2\gamma\tilde{\varepsilon}_x}\left[\xi_1\frac{1}{\gamma^2}\frac{a_0^2}{h^2} + \beta^2\left(\varepsilon_1\frac{a_0^2}{h^2} + \varepsilon_2\frac{a_0^2}{g^2}\right)\right],$$

(4.55)

$$\text{for non-penetrating fields.}$$

Note that the effect of images is always to defocus the beam, thus reducing the tune. For comparison, we quote the results presented by Reiser [1] for the effect of image forces in a cylindrical pipe in terms of phase advances and the beam perveance K:

$$\sigma_{\text{eff}} = \begin{cases} \sigma_0\left(1 - \dfrac{K\gamma^2}{h^2\sigma_0^2}S^2\right)^{1/2}, & \text{penetrating fields,} \\[2ex] \sigma_0\left(1 - \dfrac{K}{h^2\sigma_0^2}S^2\right)^{1/2}, & \text{non-penetrating fields.} \end{cases}$$

(4.56)

In these equations, σ_{eff} is the effective phase advance per lattice period 'S,' $\sigma_0 = k_0 S$ is the zero-current phase advance, and $k_0 = \tilde{\varepsilon}_x/a_0^2$. Recalling that tune and phase advance are connected through $\nu = (\rho_m/S)\sigma$, we can rewrite the second equation (4.56) in the form:

$$\nu_{\text{eff}} - \nu_0 \cong -\frac{K}{2\nu_0}\left(\frac{\rho_m}{h}\right)^2, \quad \text{or} \quad -\frac{K\rho_m}{2\tilde{\varepsilon}_x}\left(\frac{a_0}{h}\right)^2,$$

(4.57)

which agrees with equation (4.55) and is *always* a good approximation for small beams and small off-center displacements relative to the pipe radius. This result has been experimentally verified by Sutter *et al* [20]. Furthermore, by comparing equation (4.57) with equation (4.38) we can see that the coherent tune shift is typically much smaller than the incoherent tune shift.

From the results for incoherent and coherent SC tune shifts above, we see that at high energies the tune shift is dominated by image effects ($1/\gamma^2 \to 0$ in equation (4.53)); but at low energies, or with complete charge neutralization at high energies (see [18]), direct SC dominates.

All the results discussed so far apply to coasting, i.e., unbunched, beams. See [13], for example, for a treatment of tune shifts in bunched beams; [13] also incorporates all cases in a single 'practical' formula and presents interesting examples of ways to overcome the 'SC limit' by increasing the energy used for injection into circular machines. The 'SC limit' is often called the 'Laslett tune-shift limit [or] criterion' and put at $\Delta\nu \leq 0.5$. As mentioned before, this criterion is related to resonance crossing and not to reducing the SC intensity parameter χ. Direct SC and image forces are, of course, parts of the physics of *instabilities*, a host of phenomena of increasing importance for accelerators. Instabilities place additional limits on the operation of accelerators but, except for a brief discussion in chapter 9, reviewing them is beyond the scope of this book.

4.7 Computer resources

The trace-space plots in figure 4.1 were generated using the Mathcad programs ***PhSpDist-Emitt-Linear-Simplified.xmcd*** and ***PhSpDist-Emitt-II3-Simplified.xmcd***. The first program solves the equations of motion for a number of particles (typically 100 or less) in a linear uniform-focusing system 5 m in length and allows the trace space to be plotted at a number of locations. In addition, the program is used to calculate effective emittances at the same locations. The second program includes third-order terms (spherical aberration) in order to display the complex evolution of trace space. For both programs, all particles start with $x(0) = 0$, and have initial slopes that follow either uniform (rand:=0) or random (rand:=1) distributions. Note that the programs are started in such a way that the same *seed* is used every time to generate the random slopes; this ensures reproducibility. One interesting issue to explore is numerical convergence: for a random distribution of initial slopes, the effective emittance far from the source increases as the number of particles increases (why?).

Another Mathcad program, ***EnvEqn-SmApprox.xmcd***, is used to solve the K–V envelope equation in a periodic solenoid system with an arbitrary number of hard-edge solenoids. The program also includes uniform-focusing approximation calculations with and without SC, as shown in figure 4.2(b). The envelope for the matched beam in the periodic solenoid system (figure 4.1(a)) relies on output from the Matlab program MENV, which is used for matching calculations (see appendix A).

References

[1] Reiser M 2008 *Theory and Design of Charged Particle Beams* 2nd edn (Weinheim: Wiley-VCH)

[2] Wangler T P 2008 *RF Linear Accelerators* 2nd edn (Weinheim: Wiley-VCH)

[3] Edwards D A and Syphers M J 2004 *An Introduction to the Physics of High Energy Accelerators* (Weinheim: Wiley-VCH)

[4] Barletta, Spentzouris and Harms 2012 U.S. Particle Accelerator School Notes http://uspas.fnal.gov/materials/12MSU/emitlect.pdf

[5] Rhee M J 1986 Invariance properties of the root-mean-square emittance in a linear system *Phys. Fluids* **29** 3495

[6] Davidson R C and Qin H 2001 *Physics of Intense Particle Beams in High Energy Accelerators* (London: Imperial College Press and World Scientific)

[7] Bernal S 2015 Conceptual difficulties of a thermodynamics description of charged-particle beams *Proc. of IPAC2015* 649

[8] Lawson J D, Lapostolle P M and Gluckstern R L 1973 Emittance, entropy and information *Part. Accelerators* **5** 61–5

[9] O'Shea P G 1998 Reversible and irreversible emittance growth *Phys. Rev.* E **57** 1081

[10] Wangler T P, Crandall K R, Mills R S and Reiser M 1985 Relation between field energy and RMS emittance in intense particle beams *IEEE Trans. Nucl. Sci.* **NS-32** 2196

[11] Bernal S 2008 Emittance and emittance measurements U.S. Particle Accelerator School http://uspas.fnal.gov/materials/08UMD/Emittance%20and%20Emittance%20Measurements.pdf

[12] Wangler T P 2008 Resolve the apparent paradox that the most effective approach for controlling space-charge effects is to apply strong focusing, which makes the beam size small and increases the space-charge-force. Unpublished Technical Note, University of Maryland, College Park (See this book's website)

[13] Schindl K 2006 Space charge *CERN 2006-002* 305–20

[14] Rossbach J and Schümuser P 1994 Basic course on accelerator optics *CAS CERN* **94-01** 75–6

[15] Cornacchia M 2012 Incoherent space charge tune shifts, UMER Technical Note (unpublished). (See this book's website)

[16] Hofmann A Tune shifts from self-fields and images *CAS CERN* 329–48

[17] Bernal S, Li H, Kishek R A, Quinn B, Walter M, Reiser M and O'Shea P G 2006 RMS envelope matching of electron beams from 'zero' current to extreme space charge in a fixed lattice of short magnets *Phys. Rev. ST Accel. Beams* **9** 064202

[18] Keil E 1972 Intersecting storage rings *CERN* 1–52

[19] Bryant P J 1987 Betatron frequency shifts due to self and image fields *CAS CERN* 62–78

[20] Sutter D F, Cornacchia M, Bernal S, Beaudoin B, Kishek R A, Koeth T and O'Shea P G Current dependent tune shifts in the University of Maryland Electron Ring (UMER) *Proc. 2011 Particle Accelerator Conf. (New York)* p 1668 https://accelconf.web.cern.ch/PAC2011/papers/wep102.pdf

IOP Publishing

A Practical Introduction to Beam Physics and Particle Accelerators (Third Edition)

Santiago Bernal

Chapter 5

Beam (sigma) matrix and coupled optics

The Courant–Snyder (C–S) theory presented in chapter 3 assumes that there is no coupling between the transverse degrees of freedom, i.e., the equations for the y-motion are written by just substituting y for x in equations (3.19) and (3.20) for the x-motion. However, motion in a solenoid or skew-quadrupole field in the laboratory frame involves coupling of transverse degrees of freedom, which requires special treatment. In the first two sections of this chapter, we discuss the transfer matrices of a solenoid and skew quadrupole in the lab frame as an extension of the discussion in sections 2.2 and 2.3. In section 5.3, we introduce the beam (sigma) matrix, summarize important definitions and symmetry properties, and set the stage for the discussion of the beam Eigen-emittances of a cylindrically symmetric sigma matrix in section 5.4. In the latter section, we outline particular methods for approaching coupled optics and applying them to the beam matrix in a solenoid, presenting results from Wolski, Kim, Burov, *et al* for the sigma matrix and Eigen-emittances. We follow section 5.4 with a discussion of angular momentum and the envelope equation. In section 5.6, we discuss round-to-flat (RTF) and flat-to-round (FTR) beam transformers, which utilize skew quadrupole triplets and a solenoid to uncouple (RTF) or couple (FTR) the transverse degrees of freedom. We quote general equations derived by Edwards and others for the inverse focal length of thin quadrupoles in the triplets of RTF and FTR transformers. In the last section, 'Computer resources,' we illustrate the use of the matrix code TRACE 3-D to calculate the beam imaging performed by a solenoid–'antisolenoid' pair and conclude with an example of an FTR beam transformer with thin and thick skew quadrupoles.

5.1 Solenoid focusing revisited

We discussed solenoid focusing in chapter 2 and derived the focusing constant for axisymmetric focusing in the Larmor frame (equation (2.11)) and a corresponding transfer matrix in chapter 3 (equation (3.1)). In this chapter, we explicitly derive the transfer matrix in the laboratory frame, following the treatment used by Banford [1]. We start by considering motion in the *central* region of the solenoid, where the field is uniform and has only an axial component. A positively charged particle is moving clockwise in the presence of an axial field B_z; this field is perpendicular to, and its direction is out of, the plane of the circle of radius 'r' illustrated in figure 5.1.

The particle's transverse Cartesian coordinates at the initial and end points are given by $(x_1, y_1) = r(\cos\theta_1, \ \sin\theta_1), (x_2, y_2) = r[\cos(\theta_1 - \theta), \ \sin(\theta_1 - \theta)]$, respectively, with $\theta = \theta_1 - \theta_2 > 0$. Therefore, $x_2 - x_1 = r[\cos\theta_1 \cos\theta + \sin\theta_1 \sin\theta - \cos\theta_1]$. However, $x'_1 = -r\theta'_1 \sin\theta_1$, $y'_1 = r\theta'_1 \cos\theta_1$, where the prime symbol denotes the derivative with respect to 'z.' Furthermore, if the *axial* distance between the start and end points is 'l,' which can be regarded as the *effective length* of the solenoid (see also appendix B), then $\theta'_1 = -\theta/l$ (the '$-$' sign indicates that θ_1 is decreasing) (see figure 5.1). A similar analysis for the 'y' coordinate then leads to the following transfer matrix for the *central region* of the solenoid:

$$
R_C = \begin{bmatrix}
1 & (l/\theta)\sin\theta & 0 & (l/\theta)(1 - \cos\theta) \\
0 & \cos\theta & 0 & \sin\theta \\
0 & -(l/\theta)(1 - \cos\theta) & 1 & (l/\theta)\sin\theta \\
0 & -\sin\theta & 0 & \cos\theta
\end{bmatrix}. \tag{5.1}
$$

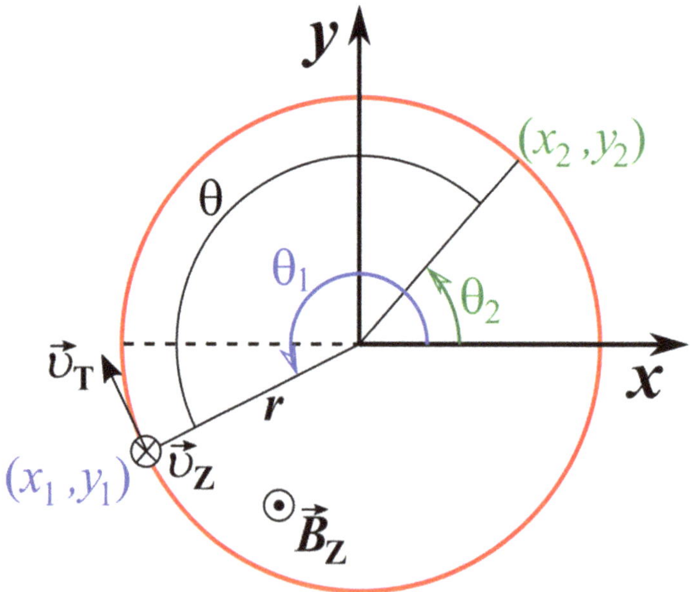

Figure 5.1. Motion inside a solenoid as seen in the lab frame—a cross section of the clockwise helical motion of a positively charged particle moving into the plane of the figure. A uniform axial *B*-field acts in the +*Z*-direction in the central region of the solenoid.

This matrix connects the vectors (x_2, x'_2, y_2, y'_2) and (x_1, x'_1, y_1, y'_1) inside the solenoid after a rotation angle θ given by the axial integration of equation (2.5):

$$\theta' C = \frac{B_z}{(B\rho)} \rightarrow \theta = \frac{B_z l}{(B\rho)}. \tag{5.2}$$

The matrix representation of focusing in the *fringe* regions of the solenoid can be obtained from the results presented in section 2.2 and illustrated by figure 2.1. We had $B_r(z) \cong -(r/2)(\partial B_z/\partial z)_{r=0}$ for the radial component of the B-field near the solenoid axis. Therefore, the Cartesian components of the B-field there are $(B_x, B_y) = -(1/2)(x, y)(dB_z/dz)_{r=0}$ and the Lorentz force equation leads to $(F_x, F_y) = \gamma m v_Z^2(x'', y'') = (1/2)q v_Z(y, -x)dB_z/dz$, if we recall that $\ddot{x} = v_Z^2 x''$, $\ddot{y} = v_Z^2 y''$. Integrating these equations from the region where $B_z = 0$ (the left-hand side of figure 2.1) to the region inside the solenoid leads to new slopes equal to

$$x'_2 = x'_1 + y_1 \frac{1}{2}\frac{B_z}{(B\rho)} = x'_1 + y_1 \frac{\theta}{2l},$$

$$y'_2 = y'_1 - x_1 \frac{1}{2}\frac{B_z}{(B\rho)} = y'_1 - x_1 \frac{\theta}{2l}, \tag{5.3}$$

where we have assumed constant (x, y), and used the definition of magnetic rigidity and equation (5.2). Finally, the matrix representation of motion through the fringe region at the solenoid entrance is

$$R_{\mathrm{FE}} = \begin{bmatrix} 1 & 0 & 0 & 0 \\ 0 & 1 & \theta/2l & 0 \\ 0 & 0 & 1 & 0 \\ -\theta/2l & 0 & 0 & 1 \end{bmatrix}. \tag{5.4}$$

The matrix for the back-end region is $R_{\mathrm{BE}} = R_{\mathrm{FE}}(-\theta)$. The overall transfer matrix of the solenoid in the laboratory frame is then $R_{\mathrm{Sol}} = R_{\mathrm{BE}}R_{\mathrm{C}}R_{\mathrm{FE}}$, or, as it is often quoted (see e.g., Banford [1], MacKay–Conte [2], or the TRACE manual [3]),

$$R_{\mathrm{Sol}} = \begin{bmatrix} R_{xx} & R_{xy} \\ R_{yx} & R_{yy} \end{bmatrix}, \quad R_{xx} = R_{yy} = \begin{bmatrix} C^2 & k^{-1}SC \\ -kSC & C^2 \end{bmatrix}, \quad R_{xy} = -R_{yx} = \begin{bmatrix} SC & k^{-1}S^2 \\ -kS^2 & SC \end{bmatrix}, \tag{5.5}$$

with $C \equiv \cos(kl)$, and $S \equiv \sin(kl)$. We have used $\theta/2l = k$, the wavenumber for oscillations in the Larmor frame, given by equation (2.10). In addition, and as also discussed by Banford [1] and others, $R_{\mathrm{Sol}} = M_{\mathrm{S}}R$, where M_{S} is the axisymmetric focusing matrix

$$M_{\mathrm{S}} = \begin{bmatrix} M_{xx} & 0_{2\times2} \\ 0_{2\times2} & M_{yy} \end{bmatrix}, \quad M_{xx} = M_{yy} = \begin{bmatrix} C & k^{-1}S \\ -kS & C \end{bmatrix}, \tag{5.6}$$

and R is the rotation matrix

$$R = \begin{bmatrix} R_{xx} & R_{xy} \\ R_{yx} & R_{yy} \end{bmatrix}, \quad R_{xx} = R_{yy} = \begin{bmatrix} C & 0 \\ 0 & C \end{bmatrix}, \quad R_{xy} = -R_{yx} = \begin{bmatrix} S & 0 \\ 0 & S \end{bmatrix}. \tag{5.7}$$

We note that the overall solenoid matrix R_{Sol} is symplectic, because it is the product of the two symplectic matrices in equations (5.6) and (5.7). However, the matrices in equations (5.1) and (5.4) are not symplectic.

To conclude, we present calculations of two-solenoid imaging with *near unit magnification* and *zero net rotation* to illustrate the theory discussed so far. The problem, originally posed by D Sutter in connection with experiments at University of Maryland, involves a 10 keV electron beam, a narrow slit, two short solenoids with opposing fields (a solenoid–'antisolenoid' pair) and a screen. The setup is illustrated in figure 5.2. We use the matrix code TRACE 3-D (see appendix A).

Since the spacing between the solenoids is comparable to their effective hard-edge lengths (6.5 cm), the calculation involves thick lenses. However, we start by using simple formulae for thin lenses to initialize the calculation in TRACE 3-D, which employs the lab-frame thick-lens matrices described earlier in this section. We model the space between the solenoids with *two drift elements* whose lengths are varied such that their sum is constant. There are, in effect, three matching variables: the first drift length between the solenoids and the *B*-fields of the solenoids.

(We encountered problems with reproducibility, which are still being explored at the time of writing, with the use of just two matching variables and a single, fixed spacing between solenoids.) Figure 5.3 displays the results for the transverse beam envelopes produced by TRACE 3-D. The thin-lens calculated *B*-fields are 121.34 G,

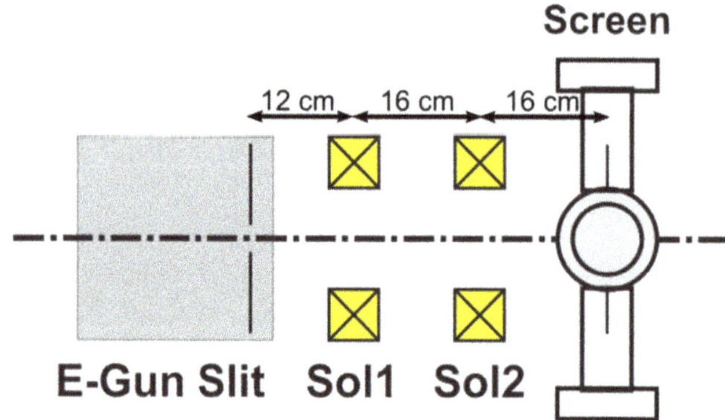

Figure 5.2. Schematics (not to scale) of solenoid–'antisolenoid' imaging of a slit.

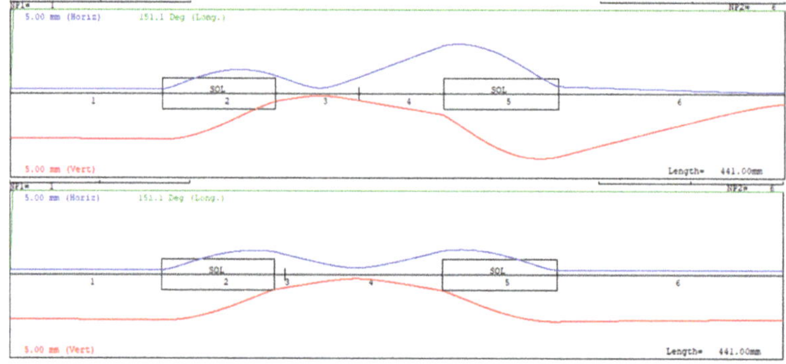

Figure 5.3. TRACE 3-D envelopes before (top) and after (bottom) optimization for unit magnification, zero net rotation imaging with a solenoid–'antisolenoid' pair.

−115.11 G, and 101.77 G, −101.06 G after optimization with thick solenoids. Using 101 G, −101 G yields exactly zero net rotation, as expected, and very close to unit magnification.

5.2 Skew quadrupole

We saw in section 2.3 that the B-field components of a *normal* quadrupole are given by

$$\left[B_x(s),\, B_y(s)\right]_{\text{Normal}} = g(s)(y,\, x), \tag{5.8}$$

where $g(s) \equiv g_x(s)$ is the field gradient. We also mention in section 2.3 and appendix B the pure 'skew' quadrupole multipole. Its B-field components can be derived from the potential $\Phi_{\text{Skew}} = c_2 r^2 \cos(2\theta) = c_2(x^2 - y^2)$, according to equation (2.13):

$$\left[B_x(s),\, B_y(s)\right]_{\text{Skew}} = g(s)(-x,\, y), \tag{5.9}$$

which is a normal quadrupole rotated by $\pi/4$, as seen in figure 5.4.

In this section, we use 'skew' to mean a normal quadrupole rotated by an arbitrary angle ψ. Under such (counterclockwise) rotation, the Cartesian coordinates transform according to $(\tilde{x},\, \tilde{y}) = (x \cos \psi + y \sin \psi,\, -x \sin \psi + y \cos \psi)$ with the

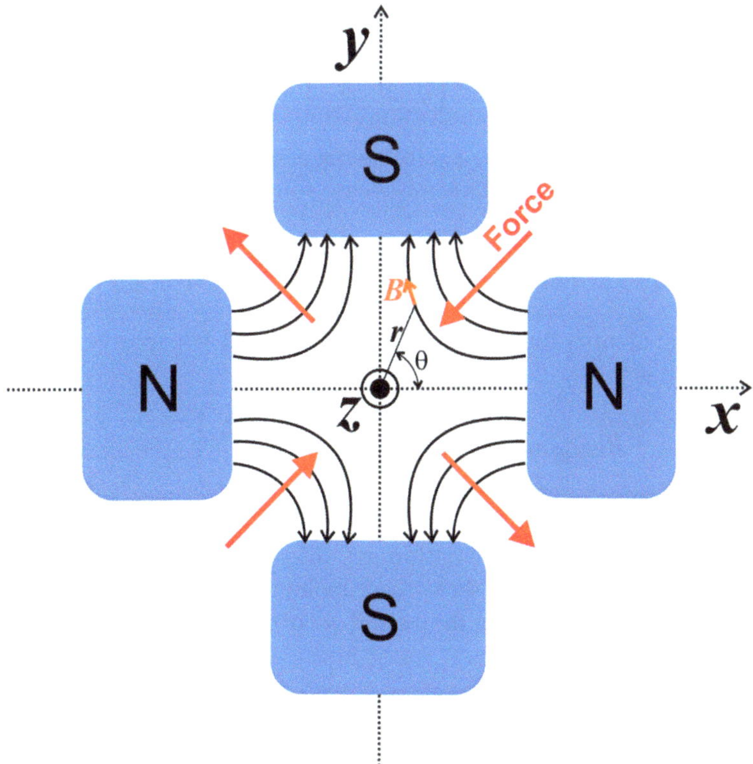

Figure 5.4. A 'pure' skew quadrupole. Compare this with figure 2.4.

inverse transformation $(x, y) = (\tilde{x} \cos \psi - \tilde{y} \sin \psi, \ \tilde{x} \sin \psi + \tilde{y} \cos \psi)$. Therefore, the original B-field components can be written in terms of the components in the rotated coordinate system, $(B_{\tilde{x}}, B_{\tilde{y}})$, as $(B_x, B_y) = (B_{\tilde{x}} \cos \psi - B_{\tilde{y}} \sin \psi, \ B_{\tilde{x}} \sin \psi + B_{\tilde{y}} \cos \psi)$, where $(B_{\tilde{x}}, B_{\tilde{y}}) = g(s)(\tilde{y}, \tilde{x})$. Finally, the original components in equation (5.8) are [4]

$$B_x(s) = g(s)[-\sin(2\psi)x + \cos(2\psi)y],$$
$$B_y(s) = g(s)[\cos(2\psi)x + \sin(2\psi)y]. \tag{5.10}$$

The corresponding single-particle equations of motion become

$$x''(s) + \kappa_x(s)[\cos(2\psi)x + \sin(2\psi)y] = 0,$$
$$y''(s) - \kappa_x(s)[\cos(2\psi)y - \sin(2\psi)x] = 0. \tag{5.11}$$

The focusing function $\kappa_x(s)$ is given in equation (2.21). Note that we recover equations (2.20) for $\psi = 0$. Equations (5.11) describe the effects of a normal quadrupole with strength $\kappa_x(s)\cos(2\psi)$ superimposed on a skew quadrupole at $45°$ with strength $\kappa_x(s)\sin(2\psi)$. The ratio of the two strengths is $\tan(2\psi) \approx 2\psi$ for small ψ. Thus, ψ must be smaller than about $0.3°$ if this ratio is to be less than 1%, which is amply satisfied in most accelerators.

To conclude this section, we present the transformation of the quadrupole matrices R_F and R_D, equations (3.9) and (3.10). Paralleling the presentation in [5], the quadrupole matrices can be incorporated into a single *block-diagonal* 4×4 matrix R_Q:

$$R_Q = \begin{bmatrix} R_F & 0_{2\times2} \\ 0_{2\times2} & R_D \end{bmatrix}. \tag{5.12}$$

This matrix transforms into R_{QRot} following a *similarity* transformation (see, for example, Arfken [6]) under a *counterclockwise* rotation $R(\psi)$ by an angle ψ:

$$R_{QRot} = R(\psi)R_Q R^{-1}(\psi), \quad R(\psi) = \begin{bmatrix} I \cos(\psi) & I \sin(\psi) \\ -I \sin(\psi) & I \cos(\psi) \end{bmatrix}, \tag{5.13}$$

where I is the 2×2 identity matrix. If $\psi = \pi/4$, the rotated 4×4 matrix representing the skew quadrupole is

$$R_{SkewQ} = \frac{1}{2} \begin{bmatrix} R_F + R_D & -R_F + R_D \\ -R_F + R_D & R_F + R_D \end{bmatrix}. \tag{5.14}$$

Skew quadrupoles are commonly used as corrector elements or in special lattices or configurations such as the beam transformers discussed in section 5.6. Wille [7], for example, discusses the use of four skew quadrupoles to compensate for the effect of the solenoid field at a detector. We defer the general discussion of coupled optics to section 5.4.

5.3 Beam (sigma) matrix

The C–S theory (section 3.3) can be complemented by a matrix description of beam second moments and their evolution in transport beamlines or periodic lattices. Let us define the sigma or covariance matrix for, say, the x-plane, as follows:

$$\mathbf{\Sigma}_{xx} \equiv \varepsilon_{x\mathrm{rms}} \begin{bmatrix} \beta_x & -\alpha_x \\ -\alpha_x & \gamma_x \end{bmatrix}, \tag{5.15}$$

where α_x, β_x, and γ_x are C–S parameters and $\varepsilon_{x\mathrm{rms}}$ is the root-mean-square (rms) emittance. Therefore, the rms beam ellipse $\gamma_x x^2 + 2\alpha_x xx' + \beta_x x'^2 = \varepsilon_{x\mathrm{rms}}$ can be written as

$$\mathbf{X}^T \mathbf{\Sigma}_{xx}^{-1} \mathbf{X} = 1, \quad \mathbf{X} \equiv \begin{bmatrix} x \\ x' \end{bmatrix}, \quad \text{with } \mathbf{\Sigma}_{xx}^{-1} = \varepsilon_{x\mathrm{rms}}^{-1} \begin{bmatrix} \gamma_x & \alpha_x \\ \alpha_x & \beta_x \end{bmatrix}. \tag{5.16}$$

Furthermore, the condition $\beta_x \gamma_x - \alpha_x^2 = 1$ is equivalent to $|\mathbf{\Sigma}_{xx}| = \varepsilon_{x\mathrm{rms}}^2$. If $\mathbf{R}$ represents the *transfer matrix* in a beamline between planes 1 and 2 with axial coordinates z_1 and z_2, the particle coordinates transform according to

$$\mathbf{X}(z_2) = \mathbf{R}\mathbf{X}(z_1) \quad \text{or} \quad \mathbf{X}(z_1) = \mathbf{R}^{-1}\mathbf{X}(z_2). \tag{5.17}$$

The sigma matrix now transforms according to (a detailed derivation can be found in [3])

$$\mathbf{\Sigma}_{xx}(z_2) = \mathbf{R}\mathbf{\Sigma}_{xx}(z_1)\mathbf{R}^T. \tag{5.18}$$

The complete *transverse* sigma matrix is often written in the form

$$\mathbf{\Sigma} = \begin{bmatrix} \langle x^2 \rangle & \langle xx' \rangle & \langle xy \rangle & \langle xy' \rangle \\ \langle x'x \rangle & \langle x'^2 \rangle & \langle x'y \rangle & \langle x'y' \rangle \\ \langle yx \rangle & \langle yx' \rangle & \langle y^2 \rangle & \langle yy' \rangle \\ \langle y'x \rangle & \langle y'x' \rangle & \langle y'y \rangle & \langle y'^2 \rangle \end{bmatrix} = \begin{bmatrix} \langle \mathbf{X}\mathbf{X}^T \rangle & \langle \mathbf{X}\mathbf{Y}^T \rangle \\ \langle \mathbf{Y}\mathbf{X}^T \rangle & \langle \mathbf{Y}\mathbf{Y}^T \rangle \end{bmatrix}, \quad \text{with } \mathbf{X},$$

$$\mathbf{Y} \equiv \begin{bmatrix} x \\ x' \end{bmatrix}, \quad \begin{bmatrix} y \\ y' \end{bmatrix}. \tag{5.19}$$

The angular brackets denote the average relative to an effective beam distribution $f(x,x',y,y')$ such as the K–V or Gaussian distributions (see chapter 4), and $\mathbf{X}\mathbf{X}^T$, $\mathbf{Y}\mathbf{Y}^T$ denote the *outer products* of vectors $\mathbf{X}$ (column), $\mathbf{X}^T$ (row), $\mathbf{Y}$, and $\mathbf{Y}^T$. As an example,

$$\langle xx' \rangle \equiv \int xx' f(x, x', y, y') dx dx' dy dy'. \tag{5.20}$$

Furthermore, note that $\mathbf{\Sigma}_{xx}$ in equation (5.15) is $\mathbf{\Sigma}_{xx} \equiv \langle \mathbf{X}\mathbf{X}^T \rangle$, and therefore

$$\langle x^2 \rangle = \beta_x \varepsilon_{x\mathrm{rms}}, \quad \langle xx' \rangle = -\alpha_x \varepsilon_{x\mathrm{rms}}. \tag{5.21}$$

A nonzero correlation term $\langle xx' \rangle$ does not indicate 'coupled' optics in the sense discussed in the first two sections, but simply a tilted C–S ellipse, as in figure 3.6.

Clearly, the 4×4 transverse sigma matrix has ten independent elements, since it is symmetrical. In addition, it can be shown [8, 9] that a beam with cylindrical symmetry is represented by a rotationally invariant sigma matrix whose 2×2 block elements satisfy $\langle \mathbf{XX}^T \rangle = \langle \mathbf{YY}^T \rangle$, $\langle \mathbf{XY}^T \rangle = -\langle \mathbf{YX}^T \rangle$, and that the off-diagonal blocks are antisymmetric, i.e., $\langle \mathbf{XY}^T \rangle^T = -\langle \mathbf{XY}^T \rangle$.

To facilitate the discussion of coupled optics in the next section, it is convenient to define a quantity $\mathcal{L}$ with dimension of length as the emittance,

$$\mathcal{L} = \langle xy' \rangle = -\langle yx' \rangle. \tag{5.22}$$

A nonzero $\mathcal{L}$ is indicative of coupled dynamics. This quantity is simply related to the average *normalized* kinetic angular momentum $\langle L \rangle$:

$$\frac{\langle L \rangle}{p_z} = \langle xy' \rangle - \langle yx' \rangle = 2\mathcal{L}. \tag{5.23}$$

Thus, using the symmetry considerations given above, the cylindrically symmetric sigma matrix can be written in the form

$$\mathbf{\Sigma} = \begin{bmatrix} \mathbf{\Sigma}_{xx} & \mathcal{L}\mathbf{J}_2 \\ -\mathcal{L}\mathbf{J}_2 & \mathbf{\Sigma}_{xx} \end{bmatrix}, \;\; with \; \mathbf{J}_2 = \begin{bmatrix} 0 & 1 \\ -1 & 0 \end{bmatrix}, \tag{5.24}$$

where $\mathbf{\Sigma}_{xx}$ is given by equation (5.15). The sigma matrix is now determined by just four quantities: two C–S parameters, the rms x-emittance $\varepsilon_{x\mathrm{rms}}$, and $\mathcal{L}$.

5.4 Coupled optics

Several different approaches are possible for the general treatment of beam transport systems that are coupled by solenoid or skew quadrupole fields. For instance, Lebedev and Burov discuss two equivalent approaches in 2D that were originated by Mais–Ripken and Edwards–Teng [10]. In this section, we outline methods advocated by Wolski [11, 12] and Kim [8, 9] for treating coupling in a solenoid field. Burov *et al* had previously discussed the same basic ideas presented in Kim's paper but in the framework of 'circular modes' [13].

Wolski generalizes the Courant–Snyder formalism, as expressed in equation (5.15), by introducing special matrices $\boldsymbol{B}^k$ such that $\mathbf{\Sigma} = \sum_k \mathbf{B}^k \varepsilon_k$. Furthermore, the two *Eigen-emittances* ε_k ($k = $ I, II) associated with the transverse 4×4 beam matrix $\mathbf{\Sigma}$ are simply related to the eigenvalues of $\mathbf{\Sigma J}$:

$$\text{eigenvalues } (\mathbf{\Sigma J}) = \pm \, i\varepsilon_k, \;\; \text{where } \mathbf{J} = \begin{bmatrix} \mathbf{J}_2 & 0 \\ 0 & \mathbf{J}_2 \end{bmatrix}. \tag{5.25}$$

$\mathbf{J}_2$ is given in equation (5.24) (actually, Wolski's analysis is more general, as it includes the full 6D phase space, revisiting linear dispersion (section 3.5) as a special case of transverse-longitudinal coupling brought about by a momentum error [12]). Furthermore, if $\boldsymbol{R}$ is the corresponding transfer matrix, its eigenvectors are also eigenvectors of $\mathbf{\Sigma J}$; in addition, if the transport system is periodic, the eigenvalues of $\boldsymbol{R}$ are related to the betatron tunes introduced in chapter 3.

Both Wolski [12] and Kim [8] consider a round beam that starts at a cathode in the presence of a uniform solenoid field B_0. This axial field can be obtained from the vector potential mentioned in equation (2.7). In Cartesian components, we have

$$(A_x, A_y) = A_\theta(- \sin \theta, \cos \theta) = \frac{B_0}{2}(-y, x). \tag{5.26}$$

Wolski derives the sigma matrix 'after acceleration to the anode' by first writing the *canonical* trace-space coordinates in the form

$$(x, y) = \sqrt{\varepsilon_{\mathrm{rms}}\beta}\left(\cos \phi_x, \cos \phi_y\right), \quad \text{and} \tag{5.27}$$

$$(x', y') = [x'_0 - (qB_0/2p_0)y, y'_0 + (qB_0/2p_0)x]. \tag{5.28}$$

We have used equation (3.19) with $\phi_x = \psi(s) + \phi$ and assumed that $\varepsilon_{x\mathrm{rms}} = \varepsilon_{y\mathrm{rms}} \equiv \varepsilon$, $\beta_x = \beta_y = \beta$, and $\alpha_x = \alpha_y = 0$ (from equation (3.21)). Furthermore, the canonical linear momenta are divided by the nominal axial momentum p_0 to obtain the dimensionless slopes x', y'. The result after averaging over the products in equation (5.19) is

$$\Sigma = \begin{bmatrix} \beta\varepsilon & 0 & 0 & \mathcal{L} \\ 0 & [1 + (\mathcal{L}/\varepsilon)^2](\varepsilon/\beta) & -\mathcal{L} & 0 \\ 0 & -\mathcal{L} & \beta\varepsilon & 0 \\ \mathcal{L} & 0 & 0 & [1 + (\mathcal{L}/\varepsilon)^2(\varepsilon/\beta)] \end{bmatrix}, \tag{5.29}$$

where all quantities are defined above. Note that all the symmetry properties of Σ mentioned above are satisfied and that the sigma matrix now depends on three parameters only: β, ε, and $\mathcal{L}$.

By contrast, Kim [8] derives the sigma matrix 'at $s = 0$,' i.e., at the cathode surface, by first writing the phase-space coordinates in the lab frame in terms of the coordinates in the rotating Larmor frame (see section 2.2). Kim's sigma matrix agrees with Wolski's result, but there is a typo in the original publication (the incorrect sign of $\langle \mathbf{YX}^\mathrm{T} \rangle$—see equations (5.19) and (5.29)). It is noteworthy that both Wolski's approach, as embedded in equations (5.27) and (5.28), and Kim's coordinate transformation are equivalent to the slope transformation at the back end of a solenoid, equation (5.3) (applicable to the front end as well), after changing the sign of θ.

In a recent paper [14], Wolski quotes a more general result for the sigma matrix, valid when $\varepsilon_{x\mathrm{rms}} \neq \varepsilon_{y\mathrm{rms}}$, $\beta_x \neq \beta_y$. It can readily be derived by generalizing equations (5.27) and (5.28); however, we retain the simpler form in equation (5.29) for the applications discussed in the next sections. Following Wolski's method, the Eigen-emittances associated with Σ, for an axisymmetric beam in a solenoid are determined using equation (5.25):

$$\varepsilon_I = \sqrt{\varepsilon^2 + 2\mathcal{L}^2\left[1 + \sqrt{1 + (\varepsilon/\mathcal{L})^2}\,\right]}, \tag{5.30}$$

$$\varepsilon_{II} = \sqrt{\varepsilon^2 + 2\mathcal{L}^2 \left[1 - \sqrt{1 + (\varepsilon/\mathcal{L})^2} \right]}. \tag{5.31}$$

It can be shown that $\varepsilon_I \varepsilon_{II} = \varepsilon^2$, so in the limit $L/\varepsilon \gg 1$, corresponding to an *angular-momentum-dominated beam*, we obtain $\varepsilon_I \approx 2\mathcal{L}$, $\varepsilon_{II} \approx \varepsilon^2/2\mathcal{L}$, or

$$\frac{\varepsilon_I}{\varepsilon_{II}} \approx \left(\frac{2\mathcal{L}}{\varepsilon} \right)^2. \tag{5.32}$$

The Eigen-emittances derived by Kim [8, 9] are

$$\varepsilon_+ = \sqrt{\varepsilon^2 + \mathcal{L}^2} + \mathcal{L}, \quad \varepsilon_- = \sqrt{\varepsilon^2 + \mathcal{L}^2} - \mathcal{L}, \tag{5.33}$$

which can easily be shown to be equivalent to those in equations (5.30) and (5.31). Burov *et al* [13] had obtained the same Eigen-emittances; in their notation,

$$2\varepsilon_{+, -} = \sqrt{\langle r^2 \rangle \langle p^2 \rangle - \langle r p_n \rangle^2} \pm \langle r p_t \rangle, \tag{5.34}$$

where the canonical momentum (dimensionless, i.e., *normalized* to the design momentum $p_0 = p_z$, as in equation (5.28)) is $\vec{p} = p_t \hat{\theta} + p_n \hat{r}$ in terms of the tangential and normal components p_t, p_n. The equivalence of equations (5.34) and (5.33) can be established by realizing that

$$4\varepsilon^2 = 4\varepsilon_+ \varepsilon_- = \langle r^2 \rangle \langle p^2 \rangle - \langle r p_n \rangle^2 - \langle r p_t \rangle^2, \quad \mathcal{L} = \frac{\langle r p_t \rangle}{2}. \tag{5.35}$$

The last equality follows from equation (5.23).

5.5 Angular momentum and the envelope equation in solenoid

Using the envelope equations (4.25) and (4.26), which are defined for the *effective* beam radius and the *effective* emittance ($4\varepsilon_{x\text{rms}}$), we can divide by two to obtain the rms envelope equation in a solenoid:

$$X''(s) + \kappa_S X(s) - \frac{K}{4X(s)} - \frac{\varepsilon_{x\text{rms}}^2}{X^3(s)} = 0, \tag{5.36}$$

where $X(s) \equiv R(s)/2$, $\kappa_S = k_0^2 = B_S^2/4(B\rho)^2$ is the solenoid focusing constant (equation (2.11)), and K is the generalized beam perveance (equation (4.12)). Reiser [15] has added a term $(p_\theta/p_z)^2 X^{-3}(s)$ in an ad hoc fashion to account for the canonical angular momentum

$$p_\theta \equiv L_{\text{can}} = \gamma m r^2 \dot{\theta} + \frac{1}{2} q B_S r^2, \tag{5.37}$$

so that the envelope equation is now

$$X''(s) + \kappa_S X(s) - \frac{K}{4X(s)} - \left[\varepsilon_{x\text{rms}}^2 + \left(\frac{L_{\text{can}}}{p_z} \right)^2 \right] \frac{1}{X^3(s)} = 0. \tag{5.38}$$

Note that the solenoid focusing term $\kappa_S X(s)$ corresponds to the edge effect discussed in section 2.2, while the canonical angular momentum L_{can} (equation (5.37)), which is conserved, contains *both* the kinetic term (rotation) and the solenoid field *inside* the solenoid.

Equation (5.38) is of no practical use with L_{can} just given as in equation (5.37). However, in the context of the beam transformers to be discussed in the next section, we can assume that the beam is 'born' inside an axial field B_S and has no initial rotation; therefore, $L_{can} = (1/2)qB_S r^2$, where 'r' is related to the *beam* size. Furthermore, on exiting the solenoid field, the beam has an average canonical angular momentum $\langle L_{can} \rangle = \langle yp_x \rangle - \langle xp_y \rangle$, or $\langle L_{can} \rangle/p_z = \langle yx' \rangle - \langle xy' \rangle = 2\mathcal{L}$, according to equation (5.23). Therefore, the rms envelope equation (5.38) can be finally written as follows:

$$X''(s) + \kappa_S X(s) - \frac{K}{4X(s)} - [\varepsilon_{xrms}^2 + (2L)^2]\frac{1}{X^3(s)} = 0. \tag{5.39}$$

We can see that the emittance and angular-momentum terms have the same dependence on the rms beam semi-axis $X(s)$. This form of the equation has been quoted by Piot and Sun [16] with our $\mathcal{L}$ differing from theirs by a factor of two. Furthermore, Piot and Sun make a distinction between canonical-angular-momentum (CAM)-dominated beams, for which $\mathcal{L}/\varepsilon \gg 1$, and magnetized beams, for which the Larmor radius is small compared to the beam radius [16]. For example, a beam exiting a skew quadrupole triplet in an FTR transformer (see the next section) is CAM-dominated but is not magnetized.

5.6 Round-to-flat (RTF) and flat-to-round (FTR) beam adapters

RTF and FTR beam transformers are analogous to the quarter-wave plate devices of light optics. In the latter, plane-polarized light is transformed into circularly polarized light after traversing a special crystal, such as quartz. The plane of polarization is oriented at 45° relative to one of the special axes of the crystal, whose length is selected so as to introduce a 90° phase difference, i.e., a quarter wavelength [17], between the two planes of polarization. A rotated quadrupole triplet plays the role of the crystal for beams. The phase difference between betatron oscillations in the two transverse planes is 90°, resulting in a beam that is not only round (an FTR transformer), but also obeys the 'vortex' condition

$$xx' + yy' = 0 \tag{5.40}$$

The principles of the FTR and RTF triplet beam adapters were initially presented by Derbenev [18], and further developed and illustrated with examples by Burov, Danilov, Derbenev, and Nagaitsev [13, 19, 20]. Wolski [11], Kim, and a few others [8, 9, 21] have discussed examples from the viewpoint of coupled-optics theory and other approaches.

The triplet adapters for FTR/RTF transformations involve a solenoid and three, 45° skew quadrupoles. The key parameter for the design of triplet adapters is the inverse focusing constant of the solenoid (in m)—see equation (2.10),

$$k_S^{-1} = 2(B\rho)/B_S \equiv \beta_S, \tag{5.41}$$

where $(B\rho)$ is the magnetic rigidity of electrons, and B_S is the B-field of the solenoid. (Several of the early papers on beam transformers use just 'β' to indicate the parameter in equation (5.41), but we prefer our notation to avoid confusion with the parameters from C–S theory or special relativity.) The overall length of the triplet, excluding the solenoid, is roughly equal to β_S. For example, Burov and Danilov [19] describe a symmetrical FTR triplet with $\beta_S = 3.33$ m at an electron energy of 500 MeV ($B_S = 10.3$ kG); the triplet length is 2.23 m. The initial flat beam, which has an unspecified emittance ratio, is transformed into a round beam with a radius equal to $1/\sqrt{2}\times$ the half-length of the horizontal dimension of the flat beam. In another example, Burov and Nagaitsev [20] illustrate an RTF asymmetric transformer at $pc = 17$ MeV, with $\beta_S = 2.83$ m ($B_S = 0.4$ kG); the triplet length is 1.6 m.

In symmetrical triplets, the two outer quadrupoles are equidistant from the central one; furthermore, the outer quadrupoles have equal strength but opposite polarity relative to the central one. In terms of C–S parameters [13, 20], $\beta_1 = \beta_S = \beta_2$, $\alpha_1 = 0 = \alpha_2$, where the subscripts refer to the initial (1) and final values (2). In asymmetric triplets [13, 20], on the other hand, $\beta_2 = \beta_S$, $\alpha_2 = 0$, while α_1, β_1 are adjustable initial parameters needed to ensure a 90° phase advance difference between the horizontal and vertical planes. It is also worth emphasizing that practically all optics treatments in the literature on triplet transformers employ the thin-lens approximation and no space-charge effects. An exception is the recent paper by Dovlatyan *et al* from the University of Maryland [22], which employs a special set of moment equations and adjoint methods to optimize FTR triplets with thick lenses and/or space-charge effects.

Experiments to date have been limited to RTF transformers involving photo-cathodes immersed in solenoid fields to produce angular-momentum-dominated beams. These beams are then injected into a skew quadrupole triplet that cancels the vortex condition, equation (5.40), to produce a flat beam. As described above, the circular-basis Eigen-emittances in the solenoid are given by equation (5.33), with an emittance ratio as in equation (5.32) in the limit of an angular-momentum-dominated beam. In the experiments at Fermilab [23], for example, the emittance ratio achieved is of the order of 100. Naturally, the opposite process would yield an angular-momentum-dominated beam starting with a flat beam of known transverse emittance ratio.

The simplest design for an FTR or RTF transformer involves a symmetrical triplet whereby the distance 'd' between the outer thin quadrupoles Q_1, Q_3 and the central one Q_2, and the *inverse* focal lengths, $q_1 = q_3$, q_2, are all given in terms of just β_S, equation (5.41) [11]:

$$d = \frac{\beta_S}{2\sqrt{1 + \sqrt{2}}}, \quad q_1 = q_3 = -\frac{\sqrt{2} + 1}{\beta_S}, \quad q_2 = \frac{2\sqrt{2}}{\beta_S}. \tag{5.42}$$

In this design, the flat beam (for an FTR transformer) is directly incident on the first quadrupole, which is located at $z = 0$, i.e., no drift space is assumed between the input beam and the first quadrupole and between the last quadrupole and the solenoid ($s_1 = s_4 = 0$ in figure 5.5). More general formulae ascribed to Edwards

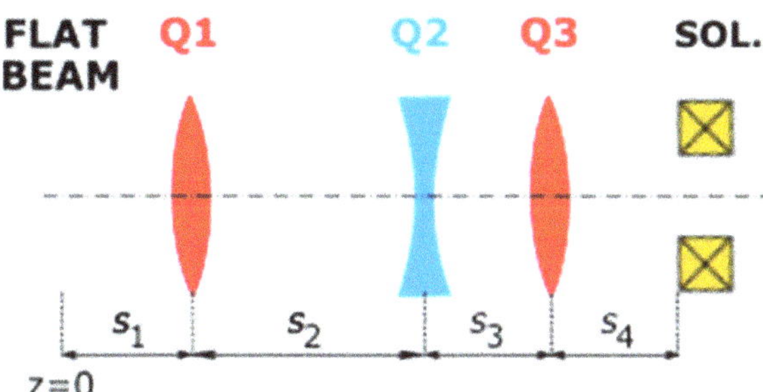

Figure 5.5. General geometry of a quadrupole triplet used for FTR beam transformation.

[9, 21] still assume thin lenses and no initial/final drift spaces, but relax the symmetry condition, i.e., (new) $s_2 \neq s_3$, $q_1 \neq q_3$.

Numerical calculations for triplet optics can rely on expressions that are more general than equation (5.42). If C is the matrix representing the skew triplet, and M, N stand for 2×2 focusing matrices in the horizontal and vertical planes, respectively, then for the un-rotated triplet, we can write [19, 20] (compare with equation (5.14))

$$\begin{bmatrix} x \\ x' \\ y \\ y' \end{bmatrix} = C \begin{bmatrix} x_0 \\ x'_0 \\ y_0 \\ y'_0 \end{bmatrix} = \frac{1}{2}\begin{bmatrix} M+N & M-N \\ M-N & M+N \end{bmatrix}\begin{bmatrix} x_0 \\ x'_0 \\ y_0 \\ y'_0 \end{bmatrix}, \tag{5.43}$$

where the column vectors represent particle coordinates in the input (subscript '0') and output planes of the skew triplet. In an RTF adapter used to produce a horizontal flat beam, we seek $(y, y') = 0$, while an FTR adapter starting with a horizontal flat beam requires $(y_0, y_0') = 0$. The general solution in either case has the form [9, 21]

$$M - N - (M+N)S = 0, \quad S = \begin{bmatrix} -\alpha_1 & \beta_1 \\ -\dfrac{1+\alpha_1^2}{\beta_1} & \alpha_1 \end{bmatrix}. \tag{5.44}$$

'S' is the 'correlation' matrix, and α_1, β_1 the initial C–S parameters of the input beam. The first equality in equation (5.44) leads to Edwards' equations [9, 21], which reduce to expressions derived by Wolski [11] in the case of a symmetric triplet and that are equivalent to equation (5.42) above.

The initial triplet design calculation can use equations (5.41), (5.32), and (5.43), starting with a choice for magnetic rigidity and solenoid field. Alternatively, it is possible to scale existing designs to the desired electron energy and solenoid field.

We now illustrate the design of an FTR adapter. For this, we use formulae from this section and the code TRACE 3-D to 'reverse engineer' an adapter originally discussed in an early Fermilab Note by Burov–Danilov [19]. We show calculations for both thin and thick quadrupoles. In the Burov–Danilov 500 MeV (electron kinetic energy) FTR transformer, the triplet geometry is *symmetric*, i.e.,

$s_1 = s_4$, $s_2 = s_3 = d$ and $q_1 = q_3$ (inverse focal lengths) in figure 5.5. Equation (5.42), the simplest design formula, does not apply exactly in this case, because $s_1, s_4 \neq 0$; however, assuming these distances are zero, we get $d = 1.072$ m for $\beta_S = 3.33$ m, which agrees with the original value for s_2. Furthermore, we obtain $q_1 = q_3 = -0.725$ m^{-1}, or $f_1 = f_3 = -1.379$ m, which is close to the quoted -1.372 m; finally, $q_2 = 0.849$ m^{-1}, or $f_2 = 1.177$ m, compared to 1.172 m. These results are reasonable, since $s_1 = s_4 = 0.045$ m $\ll s_2 = 1.072$ m. We now apply the exact formulae (5.43) and (5.44) by implementing thin-quadrupole matrices to solve for s_1, s_2, f_1, f_2. We find that a numerical solution of equation (5.44) can only be found for f_1, f_2, and that $\beta_S = 3.338$ m ($B_S = 10.00$ kG) works best to satisfy the vortex condition (5.40), which is equivalent to [19]:

$$N = -F \cdot M, \quad F = \begin{bmatrix} 0 & -\beta_S \\ 1/\beta_S & 0 \end{bmatrix}, \tag{5.45}$$

where M, N are matrices representing the un-rotated quadrupole triplet in the horizontal and vertical planes, respectively. Note that matrix S in equation (5.44) reduces to matrix F above when S is applied to a symmetrical triplet.

For completeness, we simulate the Burov–Danilov FTR triplet in TRACE 3-D; the code propagates the sigma matrix according to equation (5.18), even in the presence of rotated elements. Since the original publication [19] does not specify the parameters of the flat beam, we assume a 100:1 horizontal-to-vertical emittance ratio, or $(\varepsilon_x, \varepsilon_y) = (16, 0.16)\,\mu$m. The initial conditions (C–S parameters) are $\alpha_x = \alpha_y = 0$; $\beta_x = \beta_y = \beta_S = 3.338$ m. We first realize that using quadrupoles that have an effective length equal to the original 0.10 m does not lead to a matched beam going into the 10.00 kG solenoid; we have to use impossibly shorter quadrupoles (0.2 mm), but these are necessary to approximate the thin elements of the basic theory. Naturally, the corresponding quadrupole gradients become very large to keep the same integrated gradients of the original design. Figure 5.6 displays the

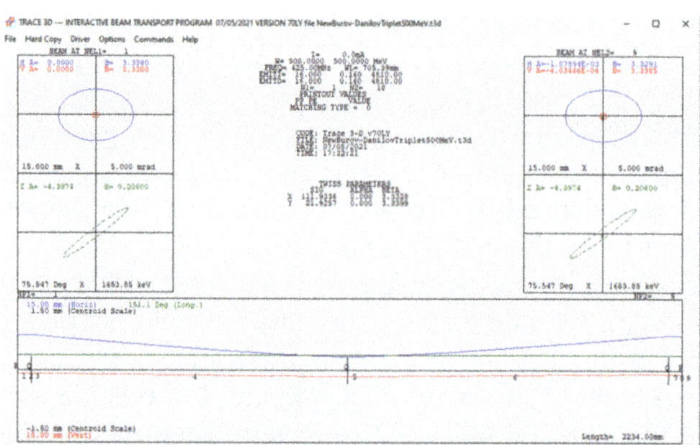

Figure 5.6. TRACE 3-D results for the initial (left panel) and final (right panel) phase spaces and transverse envelope evolution in an un-rotated thin-quadrupole Burov–Danilov FTR triplet. Notice the phase advances in X, Y.

results for the initial (left panel) and final (right panel) phase spaces and transverse envelope evolution with an un-rotated thin-quadrupole triplet. Note that the X, Y phases ('SIG') are 111.93°, 21.93°, or a 90° phase difference. Rotating the triplet by 45° leads to a round, matched beam going into the 10 kG solenoid; the beam radius is 5.2 mm, close to $1/\sqrt{2} \times 7.3$ mm, where 7.3 mm is the half-length of the horizontal dimension of the assumed initial flat beam.

In the absence of a theory for FTR/RTF transformers with thick quadrupoles, TRACE 3-D can be used to perform the optimization of the triplet parameters required to satisfy the conditions of the C–S parameters as well as the 90° phase difference between motion in the X, Y planes. The symmetric Burov–Danilov triplet can be represented using a single element in TRACE 3-D with just three adjustable quantities: the field gradient of the outer quadrupoles, the gradient of the middle quadrupole, and the distance between quadrupoles (the element has $s_1 = s_4 = 0$ by default—see figure 5.5). Thus, after setting up 0.10 m quadrupoles with the original thin-lens strengths, the optimization can proceed iteratively in two parts. First, we match the C–S parameters (matching type MT = 8) and calculate the phase advances. Second, since the phases may not satisfy the 90° difference closely enough, we match the phase advances (MT = 14) to 110°, 20° (horizontal, vertical), hoping that the C–S parameters do not become too mismatched in the process. After a few iterations, we obtain $f_1 = f_3 = -1.341\ m, f_2 = 1.130\ m, d_2 = 0.962\ m$, or $s_2 = d_2 + 0.10 = 1.062$ m. The C–S parameters are very well matched and the difference in phase advances is 90.08°. In addition, the vortex condition (5.45) and equation (5.44) are closely obeyed.

5.7 Computer resources

Explicit tests of the symplecticity of the solenoid matrices of section 5.1 are given in the Mathcad document **SolSimplTest.xmcd.** Analytical tests are provided by MacKay–Conte [2].

The Mathcad document **Sol-AntiSolThin.xmcd** contains the thin-lens imaging calculations employed in the example discussed by the end of section 5.1. The file **Sol-AntiSolThick.xmcd**, on the other hand, shows explicit calculations with the full solenoid matrices and the parameters employed in TRACE 3-D after optimization; it also includes sigma matrix calculations. The TRACE 3-D file **Sol-AntiSol-ThickInit.t3d** implements the thin-lens numbers, while **Sol-AntiSol-ThickOpt.t3d** shows the results of optimizing the imaging solution.

The Mathcad document **Burov-DanilovFNAL1998-THINQ.xmcd** contains all the relevant numeric and symbolic calculations on the symmetrical, thin-lens FTR design discussed in the previous section. (Note that some expressions must be disabled when performing symbolic algebra.) The TRACE 3-D file is **NewBurov-DanilovTriplet500MeV.t3d.**

The Mathcad document **Burov-DanilovFNAL1998-THICKQ.xmcd** has all the relevant parameters and matrix calculations for the thick-lens adapter, while the TRACE 3-D scripts **NewBurov-DanilovTriplet500MeV-Thick2.t3d** and **NewBurov-DanilovTriplet500MeV-Thick3.t3d** implement the matching ideas described above.

References

[1] Banford A P 1966 *The Transport of Charged Particle Beams* (London: E. & F. N. Spon Limited)

[2] MacKay W W and Conte M 2012 *Accelerator Physics, Example Problems with Solutions* (Singapore: World Scientific)

[3] Crandall K R and Rusthoi D P TRACE 3D documentation, LA-UR-97-886, Los Alamos National Laboratory Report, 3rd edn, May 1997.

[4] Lund S M and Barnard J J 2017 US Particle Accelerator School (USPAS) Lectures on 'Beam Physics with Intense Space Charge,' Northern Illinois University, 12–23 June 2017

[5] Rubin D 2013 Measurement and diagnosis of coupling and solenoid compensation *Handbook of Accelerator Physics and Engineering* ed A W Chao, K H Mess, M Tigner and F Zimmermann 2nd edn (Singapore: World Scientific) 4.7.4 352–6

[6] Arfken G 1970 *Mathematical Methods for Physicists* 2nd edn (New York: Academic)

[7] Wille K 1984 Skew Quad Compensation for SPEAR Minibeta Optics, SLAC/AP-27, June 1984

[8] Kim K-J 2003 Round-to-flat transformation of angular-momentum-dominated beams *Phys. Rev. ST Accel. Beams* **6** 104002

[9] Sun Y-E 2005 Angular-momentum dominated electron beams and flat beam generation *PhD Thesis* (Chicago, IL: University of Chicago)

[10] Lebedev V and Burov A 2013 Betatron motion with coupling of two degrees of freedom *Handbook of Accelerator Physics and Engineering* ed A W Chao, K H Mess, M Tigner and F Zimmermann 2nd edn (Singapore: World Scientific) 2.2.5 85–8

[11] Wolski A 2006 Alternative approach to general coupled linear optics *Phys. Rev. ST Accel. Beams* **9** 024001

[12] Wolski A 2014 *Beam Dynamics in High Energy Particle Accelerators* (London: Imperial College Press)

[13] Burov A, Nagaitsev S and Derbenev Y 2002 Circular modes, beam adapters, and their applications in beam optics *Phys. Rev. E* **66** 016503

[14] Wolski A, Christie D C, Militsyn B L, Scott D J and Kockelbergh H 2020 Transverse phase space characterization in an accelerator test facility *Phys. Rev. ST Accel. Beams* **23** 032804

[15] Reiser M 2008 *Theory and Design of Charged Particle Beams* 2nd edn (Weinheim: Wiley-VCH)

[16] Piot P and Sun Y-E 2014 Generation and dynamics of magnetized electron beams for high-energy electron cooling, Fermilab Report FERMILAB-CONF-14-142-APC June 2014

[17] Jenkins F A and White H E 1976 *Fundamentals of Optics* 4th edn (New York: McGraw-Hill)

[18] Derbenev Y 1998 *Adapting Optics for High Energy Electron Cooling* (Ann Arbor, MI: University of Michigan) UM-HE-98-04

[19] Burov A V and Danilov V V 1998 An insertion to eliminate horizontal temperature of high energy electron beam, Fermilab-TM-2043 March 1998

[20] Burov A and Nagaitsev S 2000 Courant–Snyder parameters of beam adapters, Fermilab-TM-2114 June 2000

[21] Thrane E *et al* 2002 Photoinjector production of a flat electron beam *Proc. of LINAC2002 (Gyeongju, Korea)*

[22] Dovlatyan L, Beaudoin B L, Bernal S, Haber I, Sutter D and Antonsen T M Jr 2022 Optimization of flat to round transformers with self-fields using adjoint techniques *Phys. Rev. Accel. Beams* **25** 044002

[23] Piot P, Sun Y-E and Kim K-J 2006 Photoinjector generation of a flat electron beam with transverse emittance ratio of 100 *Phys. Rev. ST Accel. Beams* **9** 031001

IOP Publishing

A Practical Introduction to Beam Physics and Particle Accelerators (Third Edition)

Santiago Bernal

Chapter 6

Longitudinal beam dynamics and radiation

The charged-particle beams treated so far were continuous, i.e., 'unbunched,' very often also referred to as 'coasting' beams. This is a good approximation whenever the longitudinal extent of the beam is long compared with its transverse dimensions. However, in most accelerators and related devices, the beam bunches have longitudinal dimensions comparable to their transverse sizes. This is the case, for example, in linear accelerators (*linacs*), where radio-frequency (RF) fields not only accelerate the beam but also impart a longitudinal density structure to it. We give a very brief introduction to RF linacs in section 6.1. In section 6.2, we describe *synchrotron oscillations* and beam bunch stability. (Synchrotron oscillations are the longitudinal counterpart of transverse betatron oscillations.) In section 6.3, we present the main ideas and basic equations related to *synchrotron radiation (SR)*. Although the topic of 'light sources' would fill volumes, we focus in section 6.4 on just their main components and insertion devices and describe the essence of free-electron lasers (FELs). Finally, in section 6.5, we introduce a number of definitions of *longitudinal beam emittance*, continue our discussion of small synchrotron oscillations, and delineate the basic model of *longitudinal space charge*: parabolic line-charge density profile and the Neuffer distribution, longitudinal beam perveance, bunch envelope equations, and the longitudinal SC intensity parameter. The 'Computer resources' section of this chapter deals with synchrotron radiation, beam bunch stability calculations, longitudinal SC, and FELs.

6.1 Radio-frequency (RF) linacs [1]

The energy of particle accelerators based on DC accelerating voltages is limited to a few tens of MeV; an example is the Van de Graaf accelerator. By contrast, RF linear accelerators, or linacs, are based on the repeated application of electric fields to achieve voltage gains that greatly exceed the maximum applied voltage. Electromagnetic waves in vacuum, however, cannot be tapped to accelerate particles, because the electric fields are perpendicular to the direction of wave propagation and are also continually changing polarity. To solve both problems, special metallic structures called *resonant cavities* are used to produce longitudinal electric fields that not only have the right polarity but also keep up with the motion of the accelerated particles.

Linacs require many RF cavities to achieve high energies because the field strengths per cavity are limited to a few MV m^{-1}. This fact explains why linacs designed for the highest energies are very long machines. In a typical electron linac, the electrons from a DC thermionic source or a photocathode gun are injected into the RF accelerating structure. The first few RF cavities capture the low-velocity electrons and accelerate them to velocities close to the speed of light; at the same time, the electrons tend to be grouped in *bunches*. The geometry of the main acceleration cavities that follow is designed to accelerate relativistic electrons. In the *traveling-wave* linac, for example, the electrons ride a traveling EM wave (whose *phase velocity* is essentially equal to c) as a surfer rides a sea wave. The cavity structure is periodic and is equivalent to a series of coupled oscillators that can sustain special field patterns called *modes* if the RF is greater than a *cutoff frequency* that depends on the cell geometry. In the *stationary-wave* type of linac, the wavelength of a longitudinal *E*-field mode is equal to an integer multiple of the iris separation; this is shown in figure 6.1 for the $2\pi/3$ mode. The wavelength in this case is equal to three times the iris separation. In the traveling-wave linac, on the other hand, the *phase advance per cell* of the RF wave is specified (e.g., $2\pi/3$ for the linac at Stanford Linear Accelerator Center (SLAC) —see chapter 8).

Ideally, all the RF power applied to a cavity would be transmitted to the beam, but this is not possible, as significant losses occur at the cavity walls. The *power loss per unit length*, P_w, is related to the *peak acceleration field E_p* through a quantity called the *shunt impedance* per unit length, r_s:

$$P_w = \frac{E_p^2}{r_s}. \tag{6.1}$$

Since the shunt impedance increases with frequency for normal conducting cavities, higher operating RF frequencies are desired. However, other important considerations that depend on the type of particle bunch needed may favor lower frequencies. Thus, the main design parameter in a linac is the frequency, which is normally chosen as a compromise between power budget considerations and the desired beam characteristics.

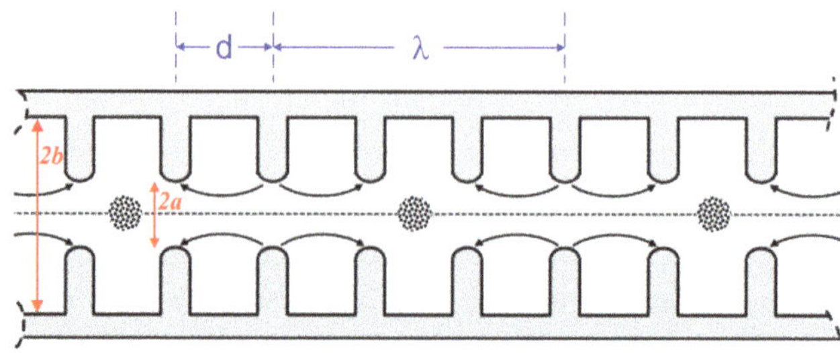

Figure 6.1. Cross section of RF cavities showing field variation of the $2\pi/3$ mode. See also section 8.1.

If W is the *stored energy* per unit length of cavity, the *transmitted energy flux* P_t (in J s^{-1}), is given by $P_t = v_g W$, where v_g is the *group velocity* of the RF wave. Furthermore, P_t decays in the axial direction with a constant 2α (units of m^{-1}), i.e., $P_t \propto \exp(-2\alpha s)$, where

$$\alpha = \frac{\omega_{\mathrm{RF}}}{2 v_g Q}.\tag{6.2}$$

In this equation, ω_{RF} is the angular RF, and Q defines the cavity's *quality factor*. In equilibrium W is constant, so the transmitted energy flux and power loss per unit length are connected through the equation

$$P_t = \frac{P_w}{2\alpha}.\tag{6.3}$$

Therefore, from equation (6.1) we get

$$E_p^2 = 2\alpha r_s P_t.\tag{6.4}$$

The peak acceleration voltage E_p decays with a constant α if the shunt impedance r_s is the same throughout the acceleration structure. The *attenuation parameter* of the *constant-impedance structure* is then equal to

$$\tau = \alpha L = \frac{\omega_{\mathrm{RF}} L}{2 v_g Q}.\tag{6.5}$$

Finally, after integration over a structure of length L, we get the *maximum energy gain per section* (see, for example, [2] for details)

$$\Delta K_{\mathrm{CI}} = e\sqrt{P_{t0} L r_s}\,\sqrt{2\tau^{-1}}[1 - \exp(-\tau)],\tag{6.6}$$

where P_{t0} is the peak RF power. The subscript 'CI' stands for 'constant impedance.'

The shunt impedance r_s is constant if all the cells in the acceleration structure have the same geometry and dimensions. Furthermore, r_s is weakly dependent on the size of the iris in the cavity cell (see figure 6.1), but the group velocity v_g and thus the transmitted energy flux P_t are very sensitive to the iris dimensions. Therefore, by making the iris diameter progressively smaller for downstream cavity cells, it is possible to achieve a constant acceleration peak field E_p. The structure so constructed is called a *constant-gradient structure*. In this case, the transmitted energy flux decreases linearly with axial distance, i.e., $\partial P_t/\partial s = \text{const.} < 0$, so the group velocity also decreases gradually. With $\tau = \alpha L$, we have (see, for example, [2])

$$v_{g\mathrm{CG}}(s) = \frac{\omega_{\mathrm{RF}}}{Q}\frac{L - [1 - \exp(-2\tau)]s}{1 - \exp(-2\tau)},\tag{6.7}$$

and a maximum energy gain per section equal to

$$\Delta K_{\mathrm{CG}} = e\sqrt{P_{t0} L r_s}\,\sqrt{1 - \exp(-2\tau)},\tag{6.8}$$

where the subscript 'CG' stands for 'constant gradient.' In section 8.1 we give a numerical example using the SLAC linac.

Other factors that affect the operation of linacs are the *beam loading*, i.e., the effect of the beam on the cavity fields, and the presence of *wakefields,* which is particularly the case for ultrarelativistic short bunches. Wakefields consist of scattered radiation that results from the interaction between bunch fields and structures in the beamline. Wakefields can, in turn, interact with parts of the bunch, leading to deleterious effects and instabilities. Chapters 10 and 11 of [3] cover these topics in detail.

6.2 Beam bunch stability and RF buckets

A crucial issue for linac operation is the beam bunch *stability*. Typically, the electron bunch is injected slightly ahead of the peak RF field. Therefore, as shown in figure 6.2, the particles experience a quasi-linear restoring force if the bunch length is small compared to the RF wavelength: the particles that arrive early at the acceleration cavity are given a smaller RF kick than those that arrive late, but those that arrive with just the right phase ϕ_S, the so-called *isochronous* particles, are given the same kick every time. Therefore, particles inside the bunch undergo longitudinal oscillations not unlike those of a pendulum; these oscillations are called *synchrotron oscillations*. The analogy with the pendulum can be carried further if we

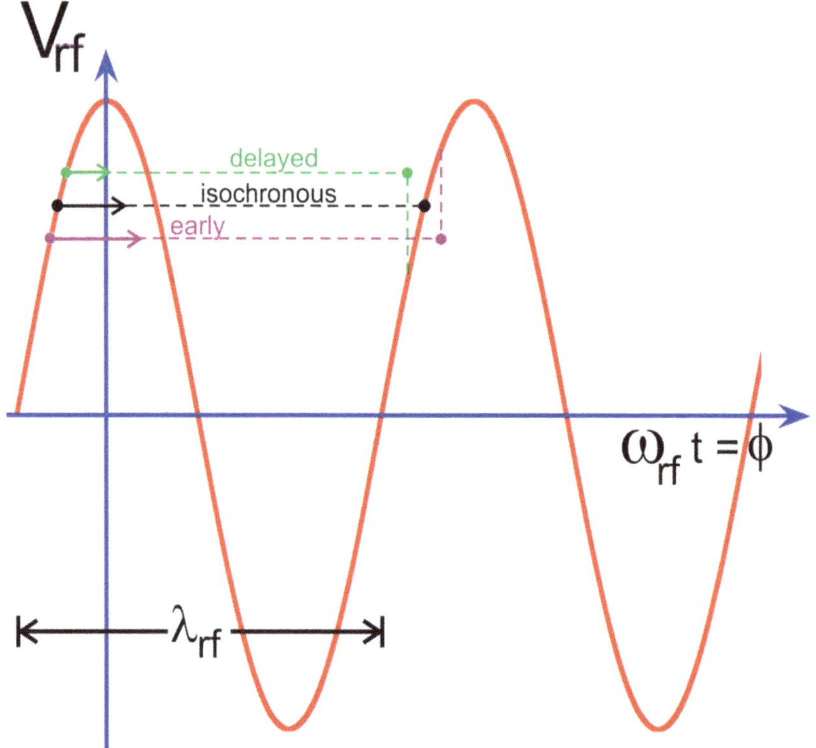

Figure 6.2. Principle of phase stability in a linac. From [1].

picture a biased pendulum, i.e., one whose equilibrium position is off from the vertical by an angle ϕ_S (see [4]). Such a pendulum is stable for oscillation angles that do not depart too much from ϕ_S. In fact, the phase space of the biased pendulum that is formed by the coordinates $(\dot{\phi}, \phi)$ is mathematically equivalent to the *longitudinal phase space* $(\delta U, \phi)$ of the bunched particles, where δU is the energy error, i.e., the energy deviation from the energy of the synchronous particles. For the pendulum we have (see section 7.5 of [4]),

$$\dot{\phi}^2 + \Omega_S^2\left(\frac{2}{\cos\phi_S}(\cos\phi_0 - \cos\phi) + 2(\phi_0 - \phi)\tan\phi_S - \dot{\phi}_0^2\right) = 0, \qquad (6.9)$$

where Ω_S is the angular frequency of *small* 'synchrotron' oscillations, and ϕ_0 is the *initial* phase angle. In figure 6.3 we have plotted equation (6.9) for $\dot{\phi}_0 = 0$, $\Omega_S = 1\ \mathrm{s}^{-1}$, $\phi_S = 60°$, and six values of ϕ_0. (Naturally, the assumed synchrotron frequency is chosen only for illustration purposes; $\Omega_S/2\pi$ in a real accelerator can be of the order of kilohertz.) The values of the initial phase ($0°$, $20°$, $30°$, $40°$, $50°$, and $120°$) correspond to six values of particle's energy; the closer the initial phase is to $\phi_S = 60°$, the closer the oscillations are to pure harmonic oscillations. If the energy (initial phase) deviates too much, on the other hand, the particle motion is unbounded, as illustrated by the open curves in figure 6.3. The stability of particles in the bunch then depends on capturing particles with the right velocities inside a region of the longitudinal RF field;

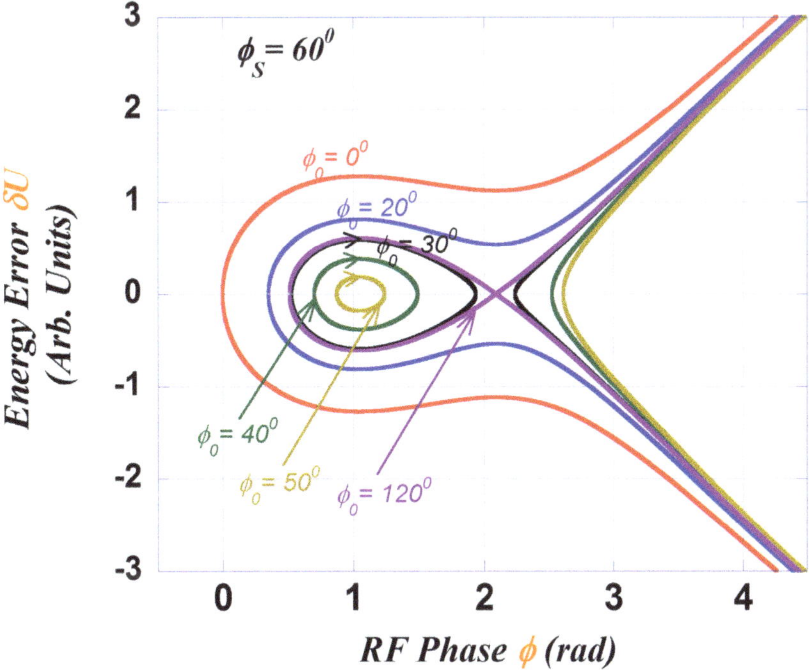

Figure 6.3. Longitudinal phase space $(\delta U, \phi)$ and RF bucket (see equation (6.9) and text). Particles inside closed curves are stable. From [1].

the corresponding region in phase space is called the *RF bucket* and is bounded by the *separatrix* curve (the curve labeled $\phi_0 = 120°$ in figure 6.3). The RF bucket contains a stable bunch of particles called a *micropulse*. If the RF is pulsed, then the group of beam bunches per pulse is called a *macropulse*. We will return to synchrotron oscillations in section 6.5.

6.3 Synchrotron radiation [1]

The force that accelerates a charged particle can be resolved into two components: one in and another perpendicular to the instantaneous velocity direction. The total power radiated by a charged particle is greater when it originates from transverse acceleration than from longitudinal acceleration by the relativistic factor γ^2:

$$P_\perp = \gamma^2 P_\parallel. \tag{6.10}$$

For highly relativistic particles, $\gamma \gg 1$; electrons with a kinetic energy of 10 MeV, for example, have $\gamma^2 = 423$. Thus, to generate copious EM radiation, it is more efficient to bend the trajectory of an energetic charged particle than to simply push it along a straight line. The radiation generated as a result of transverse acceleration is called *synchrotron radiation (SR)*.

In the *non-relativistic limit*, i.e., when the speed is a small fraction of the speed of light c, the total instantaneous power (in watts) radiated by an electron (electrical charge $-e$) with acceleration $a(t)$ is given by *Larmor's formula*:

$$P_{\text{NR}}(t) = \frac{e^2 a^2(t)}{6\pi\varepsilon_0 c^3}. \tag{6.11}$$

If relativistic effects are included, Larmor's formula becomes

$$P_R(t) = \frac{e^2 a^2(t)\gamma^4}{6\pi\varepsilon_0 c^3} = P_{\text{NR}}(t)\gamma^4. \tag{6.12}$$

If an electron of *total energy E* moves in a circular orbit of radius ρ under the action of a uniform magnetic field B, equation (6.12) can be written in other useful forms:

$$P_R(t) = \frac{e^2 c}{6\pi\varepsilon_0} \frac{\beta^4 \gamma^4}{\rho^2} = \frac{e^4}{6\pi\varepsilon_0 m_e^4 c^5} E^2 B^2. \tag{6.13}$$

Equation (6.13) displays the scaling of SR power with the bending radius ($P \propto 1/\rho^2$), rest mass ($P \propto 1/m^4$), and magnetic field of a bending dipole ($P \propto B^2$). Especially noteworthy is the very strong dependence of SR power on the rest mass of the charged particle. The energy lost to SR per turn for electrons of total energy E (in GeV) in a field B (in Tesla) can be found from equation (6.13):

$$\frac{\Delta E}{\text{turn}}[\text{keV}] = 88.5 \times \frac{E^4}{\rho} = 26.6 \times E^3 B, \tag{6.14}$$

where the bending radius ρ is given in meters. This energy must be replenished by RF acceleration cavities (see previous section) to keep the electrons circulating with the same radius ρ. From equation (6.13), we can see that the right-hand side of the first equality in equation (6.14) has to be multiplied by $(m_e/m_p)^4 = 0.88 \times 10^{-13}$ in order to be applied to protons of the same energy and orbital radius as those of the electrons. This illustrates the limitations of electron circular machines for achieving high energies. However, SR is, in itself, the reason for building electron machines as 'light sources.'

The most common SR light source is the *electron storage ring*. The main lattice components in a storage ring are dipole magnets for bending and steering, quadrupole magnets for focusing, and sextupole magnets for chromaticity corrections (see section 3.6 and appendix B). Thus, the storage ring lattice is characterized by the local radius of curvature $\rho(s)$ of the reference orbit, the quadrupole focusing function $\kappa(s)$ (equation (2.21)), the horizontal betatron function $\beta_x(s)$, and the dispersion function $D_{0x}(s)$. The momentum compaction factor, introduced in chapter 3, the energy loss per turn, and other important storage ring parameters can be expressed in terms of integrals that were introduced by Helm *et al* in 1973. The first three *SR integrals* are given by

$$I_1 = \oint \frac{D_{0x}(s)}{\rho(s)} ds, \quad I_2 = \oint \frac{ds}{\rho^2(s)}, \quad I_3 = \oint \frac{ds}{|\rho^3(s)|}. \tag{6.15}$$

The momentum compaction factor as written in equation (3.43) is then

$$\alpha_c = \frac{I_1}{C_0}, \tag{6.16}$$

where C_0 is the length of the reference orbit. The energy loss per turn is

$$U_0 = \frac{C_\gamma}{2\pi} E^4 I_2, \tag{6.17}$$

where

$$C_\gamma \equiv \frac{4\pi r_e}{3(m_e c^2)^3} \cong 8.844 \times 10^{-5} \frac{\text{m}}{\text{GeV}^3}. \tag{6.18}$$

The *nominal energy* is E, and r_e is the classical electron radius. Equation (6.17) generalizes the results of equations (6.13) and (6.14). The fourth SR integral is:

$$I_4 = \oint \frac{D_{0x}(s)}{\rho^3(s)} [1 + 2\rho^2(s)\kappa_x(s)] ds, \tag{6.19}$$

which is valid for a lattice made of sector magnets; for rectangular magnets (see section 2.5), the factor in square brackets is reduced to $2\rho^2(s)\kappa_x(s)$. The fifth SR integral is

$$I_5 = \oint \frac{H(s)}{|\rho^3(s)|} ds, \quad H(s) = \beta_x D'^2_{0x} + 2\alpha_x D_{0x} D'_{0x} + \gamma_x D^2_{0x}. \tag{6.20}$$

The fourth and fifth SR integrals are used in expressions for the *damping constants* α_i $(i = x, y, z)$, the stationary or 'natural' *rms energy spread* $(\Delta E)_{\text{rms}}$, and the *equilibrium transverse beam emittance*, ε_x, discussed next.

The emission of SR photons has important effects on the beam dynamics in storage rings and other SR sources. The first effect is the *radiation damping of betatron oscillations*. Since SR photons are emitted in a direction that is essentially identical to the direction of motion (i.e., tangential to the orbit) for highly relativistic charged particles, the conservation of momentum of the particle and the radiation leads to an overall decrease in the momentum of the particle. However, because RF replenishes only the momentum lost in the direction of the beamline, the vertical component of the momentum is reduced after acceleration. Simple considerations and algebra (see, for example, the USPAS notes by Henderson *et al* [5] or Emery [6]) yield the following equation for the vertical component of the trajectory:

$$y'' + y' \frac{1}{E_0} \frac{dE}{ds} + \kappa_y y = 0. \tag{6.21}$$

The derivation depends on equating the fractional energy and fractional momentum gains: $dE/E = dp_{\text{photon}}/p_z$, and is valid with $\beta_0 = 1$ for highly relativistic particles. Equation (6.21) has the form of a damped harmonic oscillator equation; after changing from spatial to time derivatives, we find a *damping time* and *damping constant for vertical oscillations* equal to:

$$\tau_y = \frac{2E}{U_0} T_0, \qquad \alpha_y = \frac{1}{\tau_y}, \tag{6.22}$$

where U_0 is given by equation (6.17), and T_0 is the revolution period. The damping of betatron oscillations occurs in all three dimensions and is accompanied by a reduction in the corresponding emittances. Thus, without considering other effects, the vertical emittance evolves like $\varepsilon_y = \varepsilon_{y0} \exp(-\alpha_y t)$. The damping times and constants for the components of motion in the *horizontal and longitudinal directions* can be similarly derived (see, for example, [7, 8]):

$$\tau_x = \frac{2E}{J_x U_0} T_0, \qquad \alpha_x = \frac{1}{\tau_x}, \tag{6.23}$$

$$\tau_z = \frac{2E}{J_z U_0} T_0, \qquad \alpha_z = \frac{1}{\tau_z}. \tag{6.24}$$

Equation (6.24) corresponds to the damping parameters of *synchrotron or energy oscillations*. The parameters J_i are called the *damping partition functions* and are equal to:

$$J_x = 1 - \frac{I_4}{I_2}, \qquad J_z = 2 + \frac{I_4}{I_2}, \tag{6.25}$$

where I_2 and I_4 are the SR integrals defined previously.

The damping of betatron oscillations and emittance cannot continue without limit, as it is accompanied by an 'antidamping' effect caused by random *quantum excitation*. Since the emission of SR photons also implies a reduction in energy, the particle finds itself moving very rapidly along different closed orbits. This orbital 'noise' leads to an emittance increase; however, as a result of the opposite effect caused by damping, a balance is eventually reached. The *equilibrium ('natural') horizontal emittance* and the *equilibrium ('natural') rms energy spread* are found to be (see, for example, [7, 8]):

$$\varepsilon_x = C_q \gamma^2 \frac{I_5}{I_2 - I_4}, \quad (\Delta E)_{\text{rms}} = C_q^{1/2} \gamma \left(\frac{I_3}{2I_2 + I_4} \right)^{1/2}. \tag{6.26}$$

The 'quantum constant' C_q in equation (6.26) is

$$C_q = \frac{55}{32\sqrt{3}} \frac{h}{2\pi m_e c} \cong 3.83 \times 10^{-13} \text{m}. \tag{6.27}$$

Note that the equilibrium emittance depends on the beam nominal energy through γ^2, the bending radii of the dipoles, and the lattice functions (betatron and dispersion). The rms energy spread, on the other hand, depends on the nominal energy and bending radii. In machines in which no vertical bending occurs, except possibly as a result of quadrupole misalignment or other errors, the lack of vertical dispersion reduces the antidamping effect of quantum excitation. In these machines, which comprise most electron storage rings, the vertical emittance is much smaller (by a factor of the order of 100 or more) than the horizontal emittance. Machines called 'damping rings' are designed to reduce the emittance of beams injected into *linear colliders*; a small vertical emittance can lead to luminosities that are higher by orders of magnitude (see, for example, [7, 9]).

6.4 Insertion devices and free-electron lasers (FELs) [1]

The radiation pattern of SR produced by ultrarelativistic particles forms a narrow beam cone with a vertical opening angle given by γ^{-1}. As an example, for electrons at 511 MeV the angle is one milliradian, or 3.4 min of arc. Synchrotron radiation can be obtained from bending performed by single magnets in circular machines, or using a combination of magnets of alternating polarities in *insertion devices* called *wigglers* and *undulators*. The bend angles of the individual magnets in wigglers are large compared to γ^{-1}. In contrast, the bend angles in undulators are of the same order as γ^{-1}. For bending magnet and wiggler sources, the spectrum of SR is *continuous*. Half the power is radiated above and half is radiated below a *critical photon energy* E_{cr}, which, for electrons, is given by

$$E_{\text{cr}} = \frac{3}{4\pi} \frac{hc}{\rho} \gamma^3, \quad \text{or} \quad E_{\text{cr}} \text{ [keV]} = 0.665 \times B \text{ [T]} \times E^2 \text{ [GeV]}. \tag{6.28}$$

By contrast, interference effects in undulators yield a discrete spectrum, i.e., a spectrum formed by a series of sharp peaks at certain wavelengths

$$n\lambda = \frac{\lambda_u}{2\gamma^2} \left(1 + \frac{K^2}{2} \right), \tag{6.29}$$

where $n = 1, 2, 3, \ldots$ represents the harmonic number, λ_u is the period of the undulator structure, and $K = \gamma\theta$ is the *undulator parameter*. Since the bend angle in an undulator is $\theta \approx \gamma^{-1}$, we get $K \approx 1$.

The figure of merit in many applications of SR is the *spectral brightness* or *brightness* for short (also called *brilliance* in Europe). Brightness is defined as the number of photons emitted by the SR source per unit time, per unit solid angle, per unit area at the source, and per unit bandwidth around a given frequency:

$$\text{Brightness} = \text{Photons}/(\text{s mm}^2 \text{ mrad}^2 \text{BW}) \tag{6.30}$$

Another important property of SR radiation is the degree of *coherence*. The electrons in a beam bunch emit SR due to bending performed by a simple magnet or as a result of the oscillating trajectories in an insertion device (a wiggler or an undulator). The SR emitted by individual electrons in single magnets and wigglers adds *incoherently*: if the bunch contains N particles, the total intensity (per unit frequency or *spectral bandwidth*) is simply N times the intensity produced by an individual particle. In contrast, if the electron trajectories at a given pole are all in phase, the resulting intensity of SR is proportional to N^2. This type of radiation is called *coherent synchrotron radiation* (CSR), which is seen to be produced by undulators. Although the SR produced by wigglers is incoherent, they yield more intense SR than single-bend magnets because they are longer and involve multiple bends. Furthermore, since the number of particles in a typical bunch can be very large (10^{10} for a 1.6 nC bunch, for example), the intensity of a coherent source can be many orders of magnitude larger than the intensity of an incoherent one. In reality, the SR intensity always has incoherent and coherent components; their ratio is intimately related to the structure of the beam bunch, i.e., its *line density profile*, and the geometry and dimensions of the storage ring and/or insertion device. In free-electron lasers (FELs), to be described briefly below, additional effects occur as a result of the interaction between the emitted SR and the electrons themselves.

In FELs the electron trajectories are modulated by the magnetic field of a wiggler, radiating photons that interact with the same electrons to produce *coherent bunching* accompanied by coherent radiation. Figure 6.4 shows the schematics of a basic FEL.

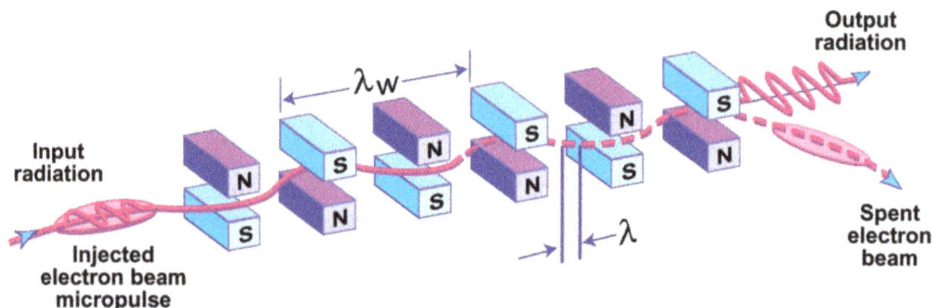

Figure 6.4. Free-electron laser concept: an electron beam interacts with a planar wiggler to produce radiation of wavelength λ. Reproduced with permission of IEEE. © [1999] IEEE. Reprinted, with permission from [10, 11].

To understand the process in more detail, we can describe the wiggler field as a traveling EM wave in the reference frame of the electron beam. This wiggler wave adds to the SR wave to produce a beat wave that has the same frequency as the SR wave but a smaller phase velocity. Because the beat wave speed is less than c, it can become synchronized with the electron beam (which has an axial speed v) and generate coherent bunches: essentially, the beat wave is an interference pattern that traps and synchronizes electron motion and radiation, leading to a process akin to stimulated emission in atomic systems. If k is the light wave number and k_w the wiggler wave number, the phase-matching or resonance condition is $\omega/(k + k_w) \approx v$. Combining this expression with the basic formula for the light wave, $\omega = ck$, we obtain the following basic relation between the FEL wavelength λ, the wiggler spacing λ_w, and the electron beam energy (as given by the gamma factor γ) for highly relativistic electrons [12]:

$$\lambda \approx \frac{\lambda_w}{2\gamma^2}. \tag{6.31}$$

A more detailed derivation shows that a factor of $(1 + K^2/2)$, as used in equation (6.29) for undulator SR at discreet wavelengths, is needed in equation (6.31). 'K,' called the FEL parameter, is of order one. From equation (6.31), it is clear that the FEL wavelength is *tunable* using the electron energy. For example, at electron energies of 4.54 and 14.35 GeV, an undulator period $\lambda_w = 3$ cm, and $K = 3.71$ (the parameters of the Linac Coherent Light Source (LCLS) at SLAC), we get $\lambda = 1.5$ nm and 0.15 nm, respectively.

FELs can be implemented as either *oscillators* or *amplifiers*. In the first case, the radiation is bounced back by mirrors, as in a regular laser cavity, until coherent radiation builds up and escapes through a semitransparent mirror. These oscillator FELs, which can be implemented in storage rings, are low-gain devices that use high-energy, low-current electron beams. In the amplifier scheme, on the other hand, laser radiation of the desired wavelength passes once through the wiggler and is amplified by the generated coherent electron bunches; this is the type of FEL called 'seeded.' Alternatively, coherent radiation can build up from noise alone in a process called self-amplified spontaneous emission (SASE). Output power saturation occurs in SASE FELs because of bunch degradation, as the electrons randomly recoil with every emitted photon. The SASE FELs (e.g., the Free Electron LASer in Hamburg (FLASH) at the German Electron Synchrotron (DESY) and LCLS at SLAC) are high-gain devices that require electron beams with very high peak current (kA) and low transverse emittance.

6.5 Longitudinal beam emittance and space charge

Since the longitudinal velocities of particles in a beam are typically orders of magnitude larger than the transverse velocities, and the energy spreads relative to the design energies are also small, the beams are considered to be 'cold' in the longitudinal direction. Thus, the Hamiltonian is separated into transverse and longitudinal components, with Boltzmann factors and corresponding temperatures defined for the transverse and longitudinal directions. This separation is, of course, artificial, as there are many instances of longitudinal–transverse coupling that tend to equalize the transverse and longitudinal kinetic temperatures over time (see [13], chapters 5 and 6). However, we will assume here that this coupling can be neglected over the temporal and spatial scales of interest.

We augment the 4D vector (x, p_x, y, p_y) of the transverse case with two additional components for the longitudinal direction: $(z, \Delta p_z/p_0)$. The longitudinal coordinate and momentum difference are: $z(t) = s(t) - s_0(t)$, $\Delta p_z(t) = p(t) - p_0(t)$. The quantities $s_0(t)$ and $p_0(t)$ represent the beam centroid position and design momentum, respectively, as measured in the laboratory frame. The *longitudinal rms emittance* (unnormalized) can be defined in terms of the rms spread in longitudinal position and rms fractional momentum deviation:

$$\varepsilon_{z\mathrm{rms}} = z_{\mathrm{rms}}\frac{(\Delta p_z)_{\mathrm{rms}}}{p_0}. \tag{6.32}$$

The normalized rms longitudinal emittance is defined as in the transverse case (chapter 4):

$$\varepsilon_{zn} = \gamma_0\beta_0\varepsilon_{z\mathrm{rms}}, \tag{6.33}$$

where γ_0, β_0 are the design values of the relativistic parameters. If we choose the total energy E and time t as canonical conjugate variables instead of z, $\Delta p_z/p_0$, the normalized emittance is defined as

$$\varepsilon_{zn}^* = (\Delta E)_{\mathrm{rms}}(\Delta t)_{\mathrm{rms}}, \tag{6.34}$$

where the asterisk distinguishes this emittance from the one in equation (6.33). In terms of the RF phase ϕ (sections 6.1 and 6.2), we can use $(\Delta\phi)_{\mathrm{rms}} = \omega_{\mathrm{RF}}(\Delta t)_{\mathrm{rms}}$ instead of $(\Delta t)_{\mathrm{rms}}$. The emittances in equations (6.33) and (6.34) are connected by the relation

$$\varepsilon_{zn} = \frac{\varepsilon_{zn}^*}{mc}, \tag{6.35}$$

as can be seen from $\Delta E = mc^2\Delta\gamma$, $\Delta p_z = mc\beta^{-1}\Delta\gamma$, $(\Delta t)_{\mathrm{rms}} = z_{\mathrm{rms}}/(\beta c)$, and $\varepsilon_{zn} = z_{\mathrm{rms}}(\Delta p_z)_{\mathrm{rms}}/mc$. (The equation for Δp_z can easily be obtained from $\Delta p_z = mc\Delta(\beta\gamma)$, and $(\beta\gamma)^2 = \gamma^2-1$.)

It is also convenient to use (z, z') as variables, where the *longitudinal angle z'* is defined by

$$z' \equiv \frac{dz}{ds_0} = \frac{\Delta\beta}{\beta_0} = \frac{\Delta p_z}{\gamma_0^2 p_0}. \tag{6.36}$$

The axial distance was defined above: $z(t) = s(t) - s_0(t)$. Equation (6.36) follows from $dz/ds_0 = (dz/dt)\,(dt/ds_0) = \Delta v_z/v_0$, and $\Delta v_z = \Delta p_z/(m\gamma_0^3)$—see equation (5.267) and problem 5–11 (part a) in [13]. The symbols Δv_z and Δp_z still stand for rms quantities and are defined in the laboratory frame. With this definition, the unnormalized rms emittance in (z, z') space, $\varepsilon_{zz'}^{\mathrm{rms}}$, is related to $\varepsilon_{z\mathrm{rms}}$(equation (6.32)) by

$$\varepsilon_{zz'}^{\text{rms}} = z_{\text{rms}} z'_{\text{rms}} = \frac{\varepsilon_{z\text{rms}}}{\gamma_0^2}. \tag{6.37}$$

The factor of γ_0^{-2} reflects the Lorentz contraction of a straight bunch. From equations (6.32), (6.37), (6.48), and $\varepsilon_{zz'} = z_m z'_m = 5\varepsilon_{zz'}^{\text{rms}}$ (see below), we can write the following for the *total longitudinal emittance*:

$$\varepsilon_{zz'} = \frac{5\varepsilon_{z\text{rms}}}{\gamma_0^2} = \frac{5z_m}{\gamma_0^2 \sqrt{5}} \frac{(\Delta p_z)_{\text{rms}}}{p_0}. \tag{6.38}$$

This can be cast in terms of the rms energy spread as

$$\varepsilon_{zz'} = \frac{5z_m}{\beta_0^2 \gamma_0^3 \sqrt{5}} \frac{(\Delta E)_{\text{rms}}}{mc^2}. \tag{6.39}$$

We have used $(\Delta p_z)_{\text{rms}}/p_0 = (\Delta E)_{\text{rms}}/(\gamma_0 \beta_0^2 mc^2)$ (see the hints following equation (6.35)). The CERN technical note by Bovet *et al* (see [14]) presents a good summary of formulas that are useful for these derivations.

We now set the stage for a simple treatment of *longitudinal space charge*. Let us assume that we have a traveling-wave linac (see section 6.1) and a beam bunch that is short compared to the RF wavelength. With this short-bunch approximation, the applied longitudinal force is *linear* (figure 6.2). The use of a traveling-wave, on the other hand, provides a physical basis for a *smooth approximation* of the actual periodic RF longitudinal focusing system.

The *applied external force* is defined relative to the synchronous particle (i.e., in the beam's frame of reference) with phase ϕ_S (section 6.2):

$$F_{az} = qE_m \left[\sin\phi - \sin\phi_s \right] \cong -\left(\frac{qE_m \omega_{\text{RF}}}{\beta_0 c} \cos\phi_s \right) z, \tag{6.40}$$

where E_m is the peak longitudinal electric field of the RF wave, $\phi - \phi_s = -(\omega_{\text{RF}}/\beta_0 c)z$, and $\phi \approx \phi_s$. Thus, the equation of motion for *small synchrotron oscillations* without space charge and with a *small rate of acceleration* can be written as [3, 4, 13]

$$z'' + \kappa_{z0} z = 0, \quad \kappa_{z0} = \frac{2\pi}{\lambda_{\text{RF}}} \frac{qE_m}{mc^2 \beta_0^3 \gamma_0^3} \left| \cos\phi_s \right|. \tag{6.41}$$

In the temporal domain, we have

$$\ddot{z} + \Omega_s^2 z = 0, \quad \Omega_s^2 = \frac{2\pi}{\lambda_{\text{RF}}} \frac{qE_m}{m\beta_0 \gamma_0^3} \left| \cos\phi_s \right|, \quad \text{straight machine}, \tag{6.42}$$

since $\Omega_s^2 = v_0^2 k_{z0}^2 = \beta_0^2 c^2 \kappa_{z0}$. A more general version of equation (6.42) that is valid for large or small oscillations leads to equation (6.9); this equation was used to plot the curves of figure 6.3 in $(\Delta E \equiv \delta U, \phi)$ space. We note that the treatment of synchrotron oscillations in [4] (the source of equation (6.9)) applies to beams in a ring accelerator, while equation (6.42) was written with a straight geometry in mind. For a beam in a ring synchrotron, equation (6.42) is replaced by

$$\ddot{z} + \Omega_s^2 z = 0, \quad \Omega_s^2 = \frac{\omega^2 h \, |\eta_{\text{tr}}|}{2\pi} \frac{qV}{mc^2 \beta_0^2 \gamma_0} \, |\cos \phi_s|, \quad \text{circular machine.} \tag{6.43}$$

The angular RF frequency is an integer multiple h (normally very large), called the *harmonic number*, of the bunch circulation frequency: $\omega_{\text{RF}} = h\omega$. Here $\omega = \beta_0 c/\rho_m$, with ρ_m = machine radius. The peak field E_m and the *peak voltage gain per turn V* are connected by $E_m = V/2\pi\rho_m$. Finally, the parameter η_{tr} (equation (3.44)) is $-1/\gamma_0^2$ for a straight accelerator because the momentum compaction factor is zero in that case (see equation (3.39)).

In addition to a linear external force, we would like the force from longitudinal space charge to be linear also. An attempt to extend the 4D Kapchinskij–Vladimirskij (K–V) distribution (chapter 4) to 6D, however, yields a nonlinear space-charge field in the longitudinal direction ([13], prob. 5–12). This extension involves a uniformly charged 6D hyperellipsoid in phase space. If we consider instead an ellipsoid in 3D with uniform volume charge density ρ_0 (in C m^{-3}), as illustrated in figure 6.5, the electrical potential can be found analytically.

If the ellipsoid shape satisfies the equation $(r/a)^2 + (z/z_m)^2 = 1$, where a, z_m are half-dimensions along r and z (figure 6.5), the *free-space potential* (i.e., with no pipe present) is quadratic in r and z, and the *line-charge density* (in C m^{-1}) inside the ellipsoid takes a parabolic profile:

$$\Lambda(z) = \Lambda_0 \left(1 - \frac{z^2}{z_m^2} \right), \quad \Lambda_0 = \rho_0 \pi a^2. \tag{6.44}$$

The longitudinal component of the self-electric field is then linear in z:

$$E_{sz}(z) = -\frac{g(z)}{4\pi\varepsilon_0 \gamma_0^2} \frac{\partial \Lambda}{\partial z} = \frac{2g(z)}{4\pi\varepsilon_0 \gamma_0^2} \frac{\Lambda_0}{z_m^2} z. \tag{6.45}$$

The function $g(z)$ defines the dimensionless *'geometry g-factor'*; it is of order unity and depends on the distance z from the bunch centroid and the aspect ratio z_m/a of the bunch. Reiser quotes results for the g-factor g_0 in free space, the g-factor at the

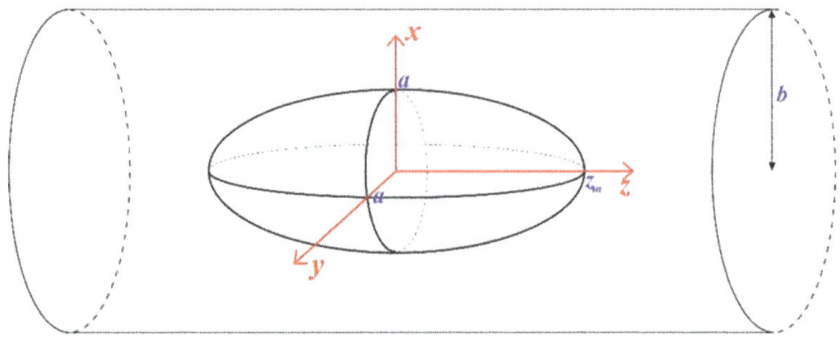

Figure 6.5. Uniformly charged ellipsoidal bunch in a cylindrical pipe.

center of the bunch $g(0)$, and the average (over the beam distribution) g-factor g that includes image effects. We only quote the asymptotic result for a *long bunch*

$$g_{\text{asymp.}}(0) = 2\ln\left(\frac{b}{a}\right), \tag{6.46}$$

where 'b' is the radius of the cylindrical pipe (figure 6.5). (A detailed derivation of equation (6.46) can be found in the USPAS notes by Barnard and Lund [15].) In reality, because of image charges, the longitudinal self-electric field is affected by the presence of the conducting pipe. Thus, deviations from linearity are significant near the ends of long bunches. Additional considerations related to the g-factor of line-charge density *perturbations* are discussed in chapter 6 of [13].

Referring again to figure 6.5, we find that the *rms bunch's length* z_{rms} is

$$z_{\text{rms}} = \langle z^2 \rangle^{1/2} = \left(\frac{\int_{-z_m}^{z_m} \Lambda(z)z^2 dz}{\int_{-z_m}^{z_m} \Lambda(z)dz} \right)^{1/2}, \tag{6.47}$$

which can easily be calculated from equation (6.44):

$$z_{\text{rms}} = \frac{z_m}{\sqrt{5}}. \tag{6.48}$$

It turns out that the line density profile of equation (6.44) corresponds to the limit of *zero longitudinal temperature* of the general line-charge density derived from the longitudinal Boltzmann factor for a matched (to the RF system) bunch. In this limit, the longitudinal space-charge force is exactly balanced by the applied *linear* external RF force (equation (6.40)). Furthermore, and as in the transverse case, the high-temperature limit or zero-longitudinal-space-charge limit yields a Gaussian density profile for $\Lambda(z)$, which is by far the most commonly used model in computer simulations of longitudinal beam dynamics.

Neuffer, in his seminal 1979 paper [16], derived the form of the particle distribution in (z, z') space that is the equivalent longitudinal version of the K–V distribution (equation (4.15)):

$$f(z, z', s) = \frac{3N}{2\pi\varepsilon_{zz'}} \sqrt{1 - \frac{z^2}{z_m^2} - \frac{z_m^2}{\varepsilon_{zz'}^2}\left(z' - \frac{z'_m}{z_m}z\right)^2}, \tag{6.49}$$

where N is the total number of particles in the bunch, z' is defined by equation (6.36), and the *total emittance* $\varepsilon_{zz'} = z_m z'_m$ satisfies the Courant–Snyder formula for the ellipse in longitudinal phase space (z, z'):

$$\gamma_z z^2 + 2\alpha_z zz' + \beta_z z'^2 = \varepsilon_{zz'}. \tag{6.50}$$

From equation (6.45) we can obtain the *focusing constant of longitudinal linear space charge*. Following the pattern of equation (6.41) we find

$$\kappa_z^{\text{sp-ch.}} = \frac{q^2 g}{2\pi\varepsilon_0 mc^2 \beta_0^3 \gamma_0^5} \frac{\Lambda_0}{z_m^2}. \tag{6.51}$$

Furthermore, since the total charge of the bunch is $Q = qN = (4/3)\Lambda_0 z_m$, and we normally have electrons, i.e., $q = -e$, we obtain $\kappa_z^{\text{sp-ch.}} = K_L/z_m^3$. K_L defines the *longitudinal beam perveance* (in m):

$$K_L = \frac{3}{2} \frac{gNr_e}{\beta_0^2 \gamma_0^5}. \tag{6.52}$$

The constant r_e is the classical electron radius. Using this notation, we can combine equations (6.41) and (6.45) to write a *single-particle* equation for longitudinal motion that includes external and internal linear forces:

$$z'' + \kappa_{z0} z - \frac{K_L}{z_m^3} z = 0. \tag{6.53}$$

The *longitudinal envelope equation*, i.e., the equation for the bunch's half-width z_m, can be derived from the Neuffer distribution, equation (6.49), and the 1D Vlasov equation (see [16, 17]), or in an ad hoc fashion by just doing the substitution $z \to z_m$ and adding an emittance term to equation (6.53):

$$z''_m + \kappa_{z0} z_m - \frac{K_L}{z_m^2} - \frac{\varepsilon_{zz'}^2}{z_m^3} = 0. \tag{6.54}$$

In terms of *rms* quantities, the envelope equation is rewritten as

$$z''_{\text{rms}} + \kappa_{z0} z_{\text{rms}} - \frac{K_L}{5\sqrt{5}\, z_{\text{rms}}^2} - \frac{(\varepsilon_{zz'}^{\text{rms}})^2}{z_{\text{rms}}^3} = 0, \tag{6.55}$$

using equation (6.48), and $\varepsilon_{zz'} = z_m z'_m = 5\varepsilon_{zz'}^{\text{rms}}$. Note that the *effective bunch half-length* is $\sqrt{5}\, z_m \cong 2.24\, z_m$, compared to the factor of $\sqrt{4} = 2$ for the transverse case (chapter 4); the corresponding effective longitudinal angle is $\sqrt{5}\, z'_m$. Another major difference between the envelope equations for the transverse and longitudinal cases is the dependence of the space-charge term on the effective beam envelope half-width. We have the space-charge term $K/2X(s)$ for the transverse case (equations (4.25) and (4.26) with $R = 2X$), but $K_L/5\sqrt{5}\, z_{\text{rms}}^2$ above.

Just as in the transverse case, a *tune depression* and a corresponding space-charge *intensity parameter* (Harris [18]) can be defined for the longitudinal beam dynamics. Furthermore, an equivalent Neuffer distribution can be defined for non-parabolic line density profiles, just as an equivalent K–V distribution is defined for arbitrary transverse particle distributions with elliptical symmetry. From the smooth approximation $z''_{\text{rms}} = 0$ of equation (6.55), we can easily obtain a longitudinal SC intensity parameter in parallel with the transverse case (equation (4.28)):

$$\chi_L = \frac{K_L/5\sqrt{5}\, z_{\text{rms}}^2}{\kappa_{z0} z_{\text{rms}}} = \frac{K_L}{K_L + \dfrac{5\sqrt{5}(\varepsilon_{zz'}^{\text{rms}})^2}{z_{\text{rms}}}}. \tag{6.56}$$

An equivalent expression [18] is $\chi_L = K_L/[K_L + (\varepsilon_{zz'}^2/z_m)]$. Beam bunches are considered to be emittance dominated if $0 \leqslant \chi_L < 0.5$ and space-charge dominated if $0.5 < \chi_L \leqslant 1$.

Because of the small energy spreads of most beams of practical interest, which imply small longitudinal emittances (equation (6.39)), longitudinal SC effects can be significant even at high energies. In particular, in the absence of longitudinal focusing, beams 'debunch,' i.e., elongate until the bunch ends meet after relatively few turns in circular machines. To illustrate this expansion, we estimate the debunching length for the 6 mA (flat-top current), 100 ns (initial-duration) bunch at the University of Maryland Electron Ring (UMER) (see chapter 9). The results of solving equation (6.54) are shown in figure 6.6. First of all, it is straightforward to show that the rms half-length of a (initially) rectangular line-charge density profile is $z_{m0}/\sqrt{3}$, where z_{m0} is the initial half-length. Thus, to construct the *equivalent parabolic profile* we need to set the equivalent maximum half-length to $z_{\mathrm{par0}} = \sqrt{5/3}\,z_{m0}$; in this way, the two profiles (figure 6.6(a)) yield the same rms half-lengths. Furthermore, the peak line-charge density of the parabolic profile must be $\Lambda_{p0} = (3/4)(Q/z_{p0})$, so that both the rectangular and parabolic bunches have the same total charge $Q = 0.6$ nC. Finally, we assume that the two bunches have the same *rms energy spread* of 100 eV, i.e., 1% of the nominal 10 keV in UMER, and use

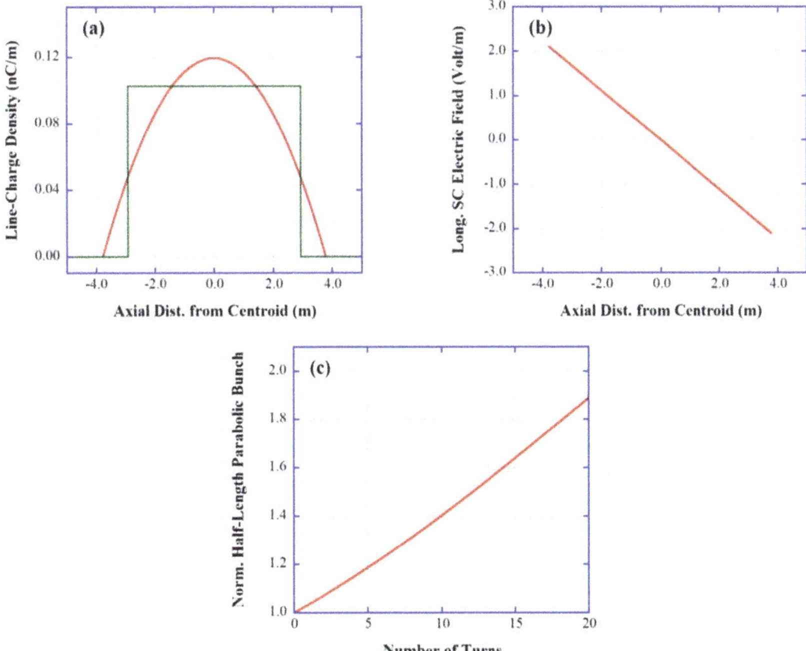

Figure 6.6. Debunching calculation for a 6 mA, 10 keV, 100 ns, 100 eV (rms energy spread) electron bunch: (a) line-charge density profiles of rectangular and (equivalent) parabolic bunches; (b) space-charge longitudinal electric field of a parabolic bunch; (c) bunch expansion vs. number of turns in UMER (see chapter 9).

equation (6.39) to calculate the total longitudinal emittance. We find that the bunch doubles its length, filling the ring, in a little more than 20 turns.

The calculation can be approached differently by taking advantage of a simple 1D fluid model (see, for example, [19]). In this model, the 'sound' (line-charge density rarefaction) speed inside the beam is

$$C_S = \sqrt{\frac{eg\Lambda_0}{4\pi\varepsilon_0\gamma_0^5 m_e}},$$

(6.57)

where g, Λ_0 are as defined previously. We find $C_S = 7.7 \times 10^5$ ms^{-1}, which implies a debunching time of 1.83 µs, or about ten turns. This latter result is closer to experimental observations [20], indicating that the fluid model may be better in this instance. The main reason for the discrepancy between the fluid and envelope models may be that the longitudinal space-charge forces at the *bunch ends* are much stronger for an initially 'square' current profile (see the first equality in equation (6.45)), as in the UMER beam, than for the parabolic profile assumed in the Neuffer model.

To conclude this section, we point out that the K–V and Neuffer models of linear space charge can be combined in an *ad hoc* fashion. Thus, a set of *coupled envelope equations* can be written ([13], section 5.4.11) which includes a redefined beam perveance containing *both* the beam radius and the bunch length. The smooth approximation form of the equations leads to algebraic equations which can be solved in cases of practical interest, such as those in which the aspect ratio of the bunch is small ('short' bunches) and the radius of the vacuum pipe is much larger than the beam radius.

6.6 Computer resources

A free program called ***Radiation2D***, written by Tsumoru Shintake of RIKEN in Japan, provides a great educational tool for illustrating the radiative pattern of a moving charged particle following different geometries. For example, the reader can study: (1) the change in electric field lines for different speeds (as fractions of light speed) of a particle moving uniformly on a straight path; the radiative patterns (again for different speeds) of a particle moving in (2) a circle; (3) a bending magnet; (4) an undulator.

The *Java* applet **FelOde**, written by Wolfgang Christian from Davidson College, very nicely illustrates a 1D pendulum model of an FEL. Pending permission, we will post the applet in the book's website. **FelOde** can be run using the simple terminal instruction 'java -cp . FelOde' in Ubuntu Linux. Especially interesting is the illustration of electron bunching in an energy-phase diagram, which is similar to the working of RF in accelerators.

We also include in this book's website a number of *Mathcad* programs: ***DipoleRadiation.xmcd*** is used to calculate the basic pattern of dipole radiation, ***BiasPend_RFBucket.xmcd*** can be employed to reproduce the results of figure 6.3, and ***LongEmitt-SpCharge.xmcd*** reproduces the calculations of bunch elongation illustrated in figure 6.6 based on the longitudinal envelope equation (6.54).

There is also the Android app TAPAs (see appendix A) which is useful for quick calculations of e.g., energy gain and RF power in linacs, synchrotron radiation parameters for electron storage rings and insertion devices, etc.

References and additional reading

[1] Joseph C L and Bernal S 2016 *Modern Devices, The Simple Physics of Sophisticated Technology* (New York: Wiley) http://bcs.wiley.com/he-bcs/Books?action=index&bcsId=10233&itemId=0470900431

[2] Wiedemann H 2007 *Particle Accelerator Physics* 3nd edn (Berlin: Springer)

[3] Wangler T P 2008 *RF Linear Accelerators* 2nd edn (Weinheim: Wiley-VCH)

[4] Conte M and MacKay W W 1991 *An Introduction to the Physics of Particle Accelerators* (Singapore: World Scientific)

[5] Henderson S, Holmes J and Zhang Y 2009 *Lecture 9a, Synchrotron Radiation*, USPAS, Nashville, January 2009

[6] Emery L 2002 *Lecture Notes Physics 575—Accelerator Physics and Technologies for Linear Colliders, winter 2002, chapter 5—Damping rings*. Argonne National Laboratory, February 7

[7] Wolski A 2014 *Beam Dynamics In high Energy Particle Accelerators* (London: Imperial College Press)

[8] Lee S Y 2004 *Accelerator Physics* 2nd edn (Singapore: World Scientific)

[9] Winick H (ed) 1995 *Synchrotron Radiation Sources—A Primer* (Singapore: World Scientific)

[10] Freund H P and Neil G R 1999 Free-electron lasers: vacuum electronic generators of coherent radiation *Proc. IEEE* **87** 782–803

[11] O'Shea P G and Freund H P 2001 Free-electron lasers: status and applications *Science* **292** 1853–8

[12] Freund H P and Antonsen T M 1996 *Principles of Free-Electron Lasers* 2nd edn (London: Chapman & Hall)

[13] Reiser M 2008 *Theory and Design of Charged Particle Beams* 2nd edn (Weinheim: Wiley-VCH)

[14] Bovet C, Gouiran R, Gumowski I and Reich K H 1970 'A selection of formulae and data useful for the design of A.G. synchrotrons', CERN Technical Note DL/70/4

[15] Barnard J and Lund S 2015 *Injectors and longitudinal physics—II*, USPAS Notes, January 19–30 Hampton, VA

[16] Neuffer D 1979 Longitudinal motion in high current ion beams: a self-consistent phase space distribution with an envelope equation *IEEE Trans. Nucl. Sci.* **NS-26** 3031–3

[17] Barnard J and Lund S 2015 *Injectors and longitudinal physics—III*, USPAS Notes, Hampton, Virginia

[18] Harris J R 2002 Longitudinal effects and focusing in space-charge dominated beams *Master's Thesis* (College Park, MD: Department of Computer and Electrical Engineering, University of Maryland) https://ireap.umd.edu/sites/ireap.umd.edu/files/documents/Harris-MS-Thesis.pdf

[19] Ho D D-M, Brandon S T and Lee E P 1991 Longitudinal beam compression for heavy-ion inertial fusion *Part. Accel.* **35** 15 http://cds.cern.ch/record/1053699?ln=en

[20] Koeth T W, Beaudoin B, Bernal S, Haber I, Kishek R A and O'Shea P G 2011 Bunch-end interpenetration during evolution to longitudinal uniformity in a space-charge-dominated storage ring' *Proc. of 2011 Particle Accelerator Conf. (New York)* https://accelconf.web.cern.ch/PAC2011/papers/moobs3.pdf

A Practical Introduction to Beam Physics and Particle Accelerators (Third Edition)

Santiago Bernal

Chapter 7

Envelope matching, resonances, and dispersion

In the first three sections of this chapter, we discuss applications or extensions of the basic theory discussed so far: *root-mean-square (rms) envelope matching* and *betatron resonances*. The first topic is a computer exercise that many accelerator designers and operators must address in practice. Charged-particle beams are born with parameters that are normally not suited for transport in the periodic lattice of a linac, ring accelerator, or storage ring. Thus, a transfer/injection section is required to modify the beam before it can be transported in the periodic lattice. We study examples of matching in periodic lattices and briefly discuss the 'source-to-cell' matching problem, including full *incoherent space charge* (SC). Furthermore, we present useful relations for calculating 'average' rms beam dimensions; these numbers can be used as initial guesses in any algorithm that finds periodic envelope solutions. In section 7.3, we present a simple treatment of *betatron resonances,* starting with integer resonances and then moving on to a more general theory. In section 7.4, we present a simple generalization of the conditions for betatron resonances that includes direct SC and discuss their connection to the beam envelope modes mentioned in the first section. Finally, in section 7.5 we consider a theory of dispersion in the presence of SC, based on the Venturini–Reiser equations. The computer resources include a *Matlab* program for *rms* envelope matching, *Mathcad* worksheets for integer resonances and envelope modes, and simulation movies for resonances. In addition, we provide a simple computer code written in C, originally from a book by MacKay and Conte on particle-accelerator problems, which can be used to explore nonlinear resonances.

7.1 Cell envelope FODO matching

Beams typically traverse a transport section that performs matching and injection into a periodic lattice. Such a lattice can be linear or closed; in the latter case, the lattice is often referred to as 'circular,' although the actual shape of the beamline is much closer to a regular polygon. In this chapter, we discuss the problem of finding periodic solutions of the Kapchinskij–Vladimirskij (K–V) envelope equations (chapter 4) in a periodic lattice. The problem of envelope matching into the periodic lattice by means of a transfer beamline is covered in books such as [1] (chapter 4) when SC is not a factor. We discuss this source-to-lattice matching, including SC, in the next section.

We consider an unbunched beam, centered on the vacuum pipe, which has a transverse distribution with elliptical symmetry. Therefore, an equivalent K–V distribution and the K–V envelope equations can be applied. We are interested in calculating the periodic beam envelopes in an asymmetric focusing–defocusing (FODO) lattice with specified zero-current phase advances per period, σ_{0X}, σ_{0Y}. We start by obtaining the quadrupole strengths that are required. The simplest calculation employs thin quadrupoles with focal lengths $f = \pm(1/\kappa_q l_q)$ and equal zero-current phase advances per period, $\sigma_{0X} = \sigma_{0Y} = \sigma_0$. We obtain, after straightforward matrix multiplication,

$$|\kappa_q| = 4\frac{\sin(\sigma_0/2)}{Sl_q} \tag{7.1}$$

The latter equation yields a peak quadrupole strength that is correct to within a few percent. However, more accurate and general results can be derived using matrices that include trajectory changes inside the quadrupoles, as well as $\sigma_{0X} \neq \sigma_{0Y}$. The relevant equations were presented in chapter 3, equations (3.12), (3.16), and (3.17), and examples given for specified zero-current phase advances. Let us start with a *symmetric* FODO lattice as shown in figure 3.2. We assume again a lattice period $S = 0.32$ m, quadrupoles with effective length $l_q = 0.0516$ m, and a zero-current phase advance $\sigma_0 = 30°$. We found by solving equation (3.12) for θ that $\kappa_q = 70.78$ m^{-2}. The beam envelope in the FODO cell depends not only on the quadrupole strength just found, but also on the beam perveance K (determined by the energy, beam current, and type of particle) and emittance. Before proceeding to numerically solve the K–V envelope equations, we can obtain a good estimate of the 'average' beam envelope transverse dimension using the smooth approximation. For an emittance-dominated beam we can employ the zero-current result $a_0 = \sqrt{\tilde{\varepsilon}_{\mathrm{rms}}/k_0}$, with $k_0 = \sqrt{\kappa_q}$. As an example, let us consider a beam with perveance $K = 1.5\times10^{-6}$ (e.g., a 0.1 mA, 10 keV electron beam) and effective emittance $\tilde{\varepsilon}_{\mathrm{rms}} = 6.0$ μm in both transverse planes. Thus, we find $a_0 = 1.9$ mm. We can use this estimate as a first guess in order to solve the K–V envelope equation (4.46) in the FODO cell. The solution for the periodic envelope is found using the periodic-matcher routine of the *Matlab* code MENV described in appendix A. The results are shown in figure 7.1. Note from the figure that the 'average' beam semi-axis dimension is 2.0 mm, very close to a_0.

The same result as that shown in figure 7.1 can be obtained using essentially any of the codes described in appendix A, but the popular matrix code TRACE 2-D would perhaps be the best choice.

The beam envelope is dramatically different if the beam is strongly SC dominated. Let us repeat the calculation for a $K = 1.5 \times 10^{-3}$ beam (e.g., a 100 mA, 10 keV electron beam) with an effective emittance of $\tilde{\varepsilon}_{\mathrm{rms}} = 60$ μm. To estimate the beam's transverse dimension, we can use $a_B = \sqrt{K/k_0}$ (see equation (4.31)), which yields $a_B = 24$ mm. The beam envelopes calculated using MENV are shown in figure 7.2.

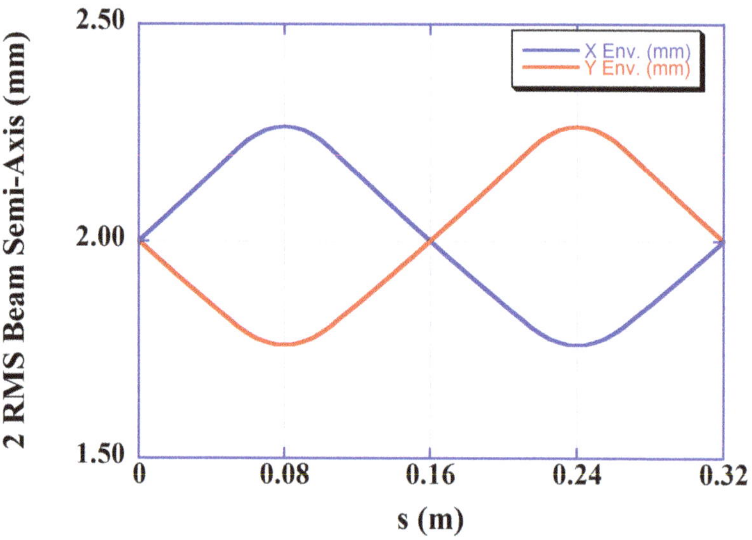

Figure 7.1. Beam envelope in a symmetric FODO lattice. The beam perveance is $K = 1.5 \times 10^{-6}$, and the effective emittance is 6.0 μm (an emittance-dominated beam). The quadrupole strengths yield a zero-current phase advance $\sigma_0 = 30°$.

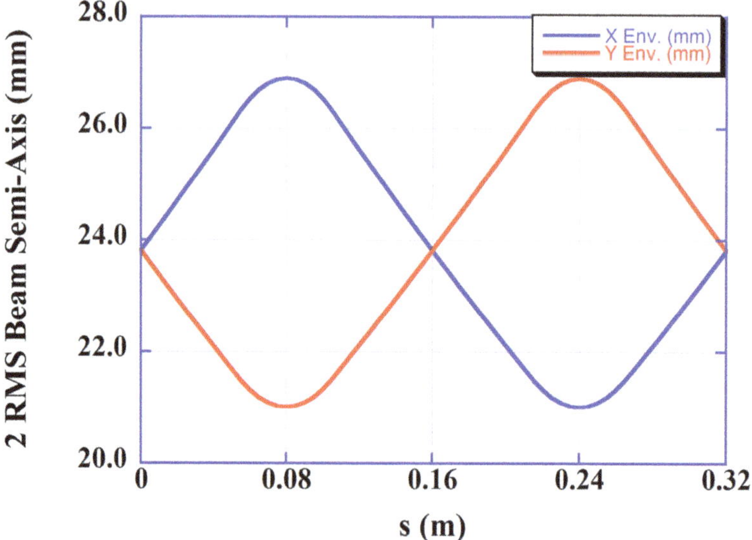

Figure 7.2. Beam envelope in a symmetric FODO lattice. The beam perveance is $K = 1.5 \times 10^{-3}$, and the effective emittance is 60 μm (a strongly SC dominated beam). The quadrupole strengths yield a zero-current phase advance $\sigma_0 = 30°$.

For intermediate SC intensities, a surprisingly good approximation for the average beam radius is provided by the relation

$$a \cong \sqrt{a_0^2(k_0, \tilde{\varepsilon}_{\text{rms}}) + a_B^2(k_0, K)}, \tag{7.2}$$

where we have explicitly indicated the dependence on the *smooth-approximation* constant k_0, the emittance, and the generalized beam perveance.

If the FODO lattice is *asymmetric* (as in figure 3.4) and $\sigma_{0X} = 30°$, $\sigma_{0Y} = 45°$, the solution for the quadrupole strengths is obtained from equations (3.16) and (3.17): $\kappa_1 = 84.67$ m^{-2}, and $\kappa_2 = 94.40$ m^{-2}. We assume the same beam perveances and emittances as those of the examples with the symmetric FODO lattice. This time, we expect different 'average' beam dimensions in the two transverse planes. If a and b represent these dimensions, the smooth-approximation equations to be solved to estimate these values are given by:

$$k_{0x}^2 a - \frac{2K}{a+b} - \frac{\tilde{\varepsilon}_x^2}{a^3} = 0,$$

$$k_{0y}^2 b - \frac{2K}{a+b} - \frac{\tilde{\varepsilon}_y^2}{b^3} = 0,$$

(7.3)

which are the same as equations (4.47). We find $(a, b) = (2.0, 1.6)$ mm for the emittance-dominated beam used previously, and $(a, b) = (27.8, 12.6)$ mm for the strongly SC dominated beam. The envelopes for both cases are shown in figure 7.3.

No bending dipole magnets were present in the previous examples. If dipoles are present, they change the FODO matching solution because of edge focusing. TRACE 2-D and 3-D are more suitable for matching in such cases.

We now present a brief discussion of the beam envelope oscillations that ensue when the beam is not perfectly matched. Starting with the K–V envelope equations (4.46), we define the *average* rms envelope semi-axis dimensions in the two transverse planes, $\overline{X}$ and $\overline{Y}$, as the solutions of the equations in the *smooth approximation*. We next assume that the $X(s)$ and $Y(s)$ envelopes deviate by a small amount from their average values. The analysis of the linearized 'small oscillations' problem is standard (see, for example, [2]); it leads to two fundamental envelope modes with *phase advances per period* given by:

$$\sigma_{\text{sym.}} = \sqrt{2\sigma_0^2 + 2\sigma^2}, \text{ symmetric mode,}$$

$$\sigma_{\text{asym.}} = \sqrt{\sigma_0^2 + 3\sigma^2}, \quad \text{antisymmetricmode.}$$

(7.4)

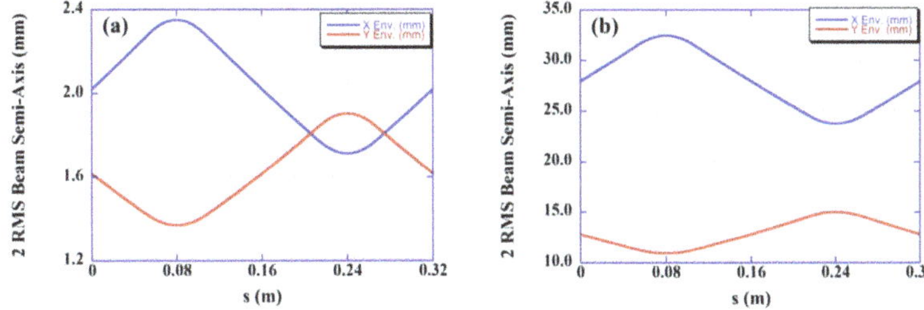

Figure 7.3. Beam envelopes in an asymmetric FODO lattice: (a) $K = 1.5 \times 10^{-6}$, 6.0 μm effective emittance; (b) $K = 1.5 \times 10^{-3}$, 60 μm effective emittance. The quadrupole strengths yield zero-current phase advances $\sigma_{0X} = 30°$, $\sigma_{0Y} = 45°$.

In these equations, σ_0, σ are the zero-current and full-current betatron phase advances per period. The first normal mode is the in-phase or 'breathing' mode; the second mode is the 'antiparallel' or quadrupolar mode. These modes are analogous to those that occur in two coupled harmonic oscillators, such as pendula. In the first mode, the pendula swing in phase, while in the second mode the phases are 180° apart. The general motion of the oscillators is a linear superposition of the two modes. With zero current the two modes become one with a phase advance per period $\sigma_{\mathrm{env}} = 2\sigma_0$, i.e., the envelope of the mismatched beam oscillates at twice the betatron frequency.

The analysis just presented has been extended to piecewise, hard-edge solenoid and quadrupole periodic lattices by Struckmeier and Reiser [2], and independently, using the K–V distribution and the Vlasov equation, by Hofmann et al [3]. The results of the analysis differ from those of the smooth approximation, especially for $\sigma_0 > 90°$, when the envelope mode oscillations have a phase advance of 180° per lattice period. This condition leads to an envelope instability or 'parametric resonance' and can happen in both linear and ring lattice geometries (see also section 7.4). The departure from the smooth-approximation results is especially large for quadrupole lattices, high σ_0, and high SC intensity, i.e., low values of the depressed phase advance σ per period.

In more recent work [4, 5], Hofmann extended the mode analysis to 2D anisotropic coasting beams using a generalized K–V distribution. The anisotropy originates from either unequal transverse focusing strengths or emittances, or both. The SC potential acts as a perturbation, and the corresponding linearized Vlasov equation is solved to find dispersion relations for the mode frequencies. The results for the 'second-order even' modes include those characterized by the phase advances (per period) in equation (7.4). In Hofmann's notation, we have

$$\sigma_{\mathrm{fast}} = \sqrt{4 + \sigma_p^2}, \text{ symmetric (fast) mode,}$$

$$\sigma_{\mathrm{slow}} = \sqrt{4 + \frac{\sigma_p^2}{2}}, \quad \text{antisymmetric (slow) mode,} \tag{7.5}$$

where σ_{fast}, σ_{slow} are *dimensionless quantities* defined in terms of the *frequency* ratios $\omega_{\mathrm{fast}}/\nu_x$, $\omega_{\mathrm{slow}}/\nu_x$, in Hofmann's notation, and $\sigma_p^2 = 2\chi = 4\Delta\nu_x/\nu_{0x}$ (for small SC) in terms of the SC intensity parameter defined in chapter 4 (see equations (4.28) and (4.38)). The equivalency of equations (7.5) and (7.4) can be easily checked by realizing that $\nu_{0x}^2 - \nu^2 \cong 2\nu_{0x}\Delta\nu_x$ for small SC. It is important to note here that tunes —dimensionless—and frequencies can be used interchangeably as long as the units' consistency is satisfied.

A formula often quoted for computing the incoherent tune (or frequency) shift from the 'quadrupolar' (slow) mode tune (or frequency) ν_{asym} is [5]

$$\Delta\nu_x = \frac{2\nu_{0x} - \nu_{\mathrm{asym.}}}{\frac{1}{2}\left(3 - \frac{\eta_0}{1 + \eta_0}\right)}, \quad \eta_0 \equiv \frac{a}{b}, \tag{7.6}$$

where a, b are the horizontal and vertical beam dimensions in Hofmann's model. This expression *does not* follow from either equation (7.4) or (7.5) but from a more general equation valid for the case in which the horizontal and vertical tunes are 'sufficiently' split (see [5, 6]). Additional considerations are given in section 7.4.

To conclude this section, we present a numerical example of mismatched envelope modes in a simple context. We solve the K–V envelope equations (4.46) for a lattice of hard-edge quadrupoles corresponding to one turn (36 FODO periods) in the University of Maryland Electron Ring (UMER, see section 9.1). We use a 10 keV, 6.0 mA, 26.2 μm (effective emittance) electron beam. The *effective* hard-edge focusing constant (see the last section of appendix B) for a matched beam is 144.27 m^{-2}, i.e., a zero-current phase advance per period of approximately $\sigma_0 = 66°$. The first two quadrupoles are overpowered by 10% to induce an envelope mismatch. A fast Fourier transform (FFT) of the difference $(X - Y)$ and sum $(X + Y)$ of the 2×*rms* horizontal (X) and 2×*rms* vertical envelopes (Y) then reveals three main peaks corresponding to the wavelengths $\lambda_0 = 0.32$ m (a full-lattice period), $\lambda_{\mathrm{odd}} = 1.15$ m (the quadrupolar or odd mode), and $\lambda_{\mathrm{even}} = 1.05$ m (the breathing or even mode). Figure 7.4 illustrates the results. Calculations based on the smooth approximation yield λ_{odd}, $\lambda_{\mathrm{even}} = 1.18$, 1.04 m, respectively.

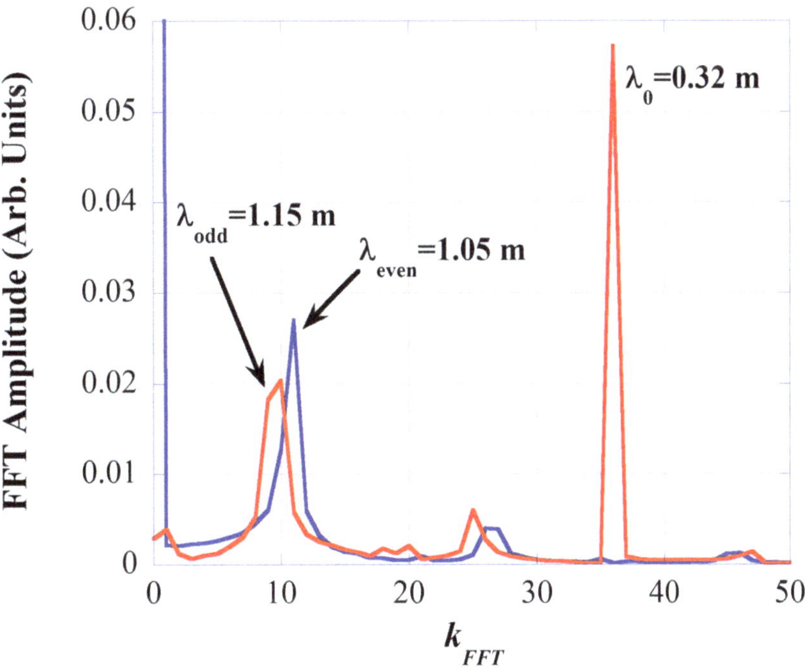

Figure 7.4. Envelope mismatch modes: the two peaks on the left represent the two mismatch oscillation modes of a 10 keV, 6.0 mA beam in an idealized UMER model. See the text and the last section of this chapter, 'Computer Resources.'

7.2 Source-to-cell envelope matching

A charged-particle beam must transition from its source to a periodic linear or ring lattice, where the beam envelope should evolve periodically as described in the previous section. Since there are two envelope dimensions and two envelope slopes to match, X, Y, X', Y', a minimum of four adjustable quadrupole lenses are needed at *fixed locations*. Reference [1] discusses *variable-geometry* configurations of quadrupoles (doublet, triplet, and quadruplet) whereby the locations of the magnets are additional parameters for matching; this additional freedom can yield matching solutions for the Courant–Snyder parameters α, β (doublet and triplet), and α, β, D—dispersion (triplet and quadruplet). However, in many practical situations, the magnets cannot easily be relocated and/or the analytical methods do not yield possible solutions for the locations of the magnets and/or their gradients. In any case, the analytical methods can serve as a first step in the design of matching sections, also called *transfer lines*.

When SC is significant, rms envelope matching is accomplished by iteratively solving the K–V envelope equations from the source to the target plane where the periodic envelope begins. The envelope codes SPOT and MENV, described in appendix A, provide the means for finding magnet parameters that yield source-to-cell rms matching. Alternatively, standard matrix codes, particularly TRACE 2-D and 3-D (see appendix A), can be employed. Frequently, because of the nature of the optimization embedded in the nonlinear control methods often used [7], there are multiple solutions for a given matching problem. Figure 7.5

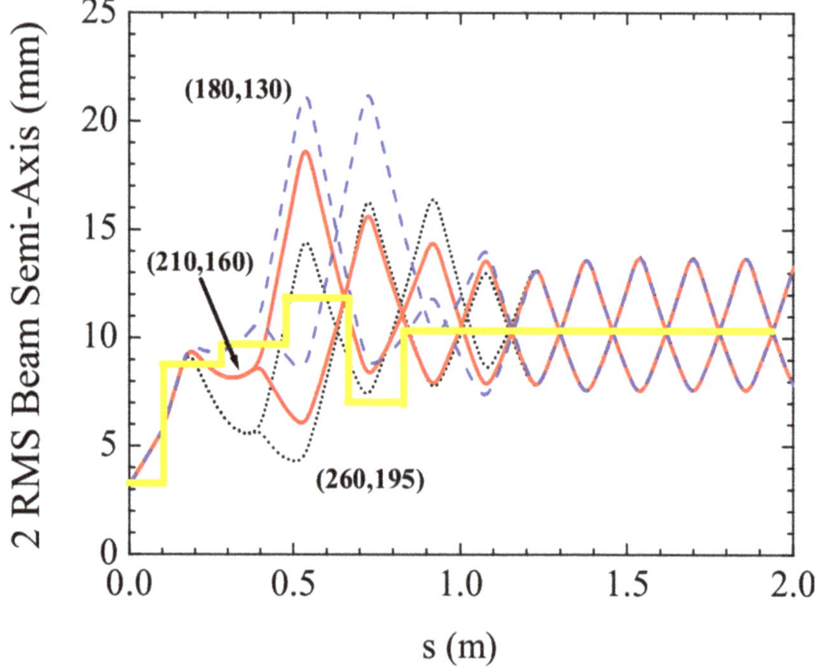

Figure 7.5. Source-to-cell rms beam envelope matching calculations in UMER with extreme SC. Three solutions for $\sigma_0 = 72.92°$ are shown (SPOT code—see appendix A—with smooth-gradient quadrupoles). The numbers in parentheses indicate the peak strengths (in m^{-2}) of a solenoid and first quadrupole. Adapted from [7].

illustrates three matching solutions for a 100 mA, 10 keV, $\varepsilon_{x,y} = 60$ µm electron beam in UMER in an early version of the matching section consisting of one short solenoid and seven air-core printed-circuit quadrupoles [7]. The envelope solutions are labeled with the peak strengths of the solenoid and first quadrupole, and the yellow trace represents a chosen 'reference trajectory,' a feature of the SPOT envelope code used to optimize the solution. The envelope trace in red is deemed to be the one with the desired smallest 'flutter.' The envelope calculations are in good agreement with experimental results [7].

7.3 Betatron resonances

The betatron resonance condition in a circular machine such as a synchrotron or storage ring is expressed by

$$n_x \nu_{x0} \pm n_y \nu_{y0} = Np, \tag{7.7}$$

where n_x, n_y, p, and N are integers, and ν_{x0} and ν_{y0} are the horizontal and vertical bare tunes (the notation Q_{x0}, Q_{y0} is used in Europe). N is the *superperiodicity* of the machine, and $|n_x| + |n_y|$ is the *order* of the resonance. When $N > 1$ and integer, we have 'structural' or 'systematic' resonances, whereas when $N = 1$, the resonances are associated with random errors in magnet strengths or locations. These latter resonances are called 'non-structural' or imperfection resonances. A *working or operating point* (ν_{x0}, ν_{y0}) is chosen so as to avoid the condition equation (7.7).

The most destructive resonances are *linear resonances*, i.e., first- and second-order resonances. First-order resonances are associated with dipole strength and rotational errors and with quadrupole displacement errors. *Second-order resonances*, on the other hand, are caused by quadrupole (i.e., gradient) errors. The names 'integer' and 'half-integer' resonances are also employed to describe first- and second-order resonances, respectively. Further, the onset of linear resonances is independent of particle amplitude, but *nonlinear* resonances, i.e., resonances of the third or higher orders, are amplitude dependent. We will give a motivation for equation (7.7) after discussing a simple treatment of integer resonances.

Dipole errors arise from bending magnet errors and from quadrupole transverse displacement errors. We now show that particle amplitude in integer resonances grows linearly by the turn. A dipole error leads to an angular error (or 'kick') given by (see chapter 3)

$$x' = \frac{\delta B}{B\rho} \Delta s, \tag{7.8}$$

where Δs is the dipole's effective length and ρ is the bend radius. Therefore, if x'_0 is the initial slope at a given point in the reference orbit, the kick leads to new slopes

$$x'_1 = x'_0 + \frac{\delta B}{B\rho} \Delta s, \quad x'_2 = x'_1 + \frac{\delta B}{B\rho} \Delta s, \quad ..., \tag{7.9}$$

after the first, second, and so on, revolutions. In general, we have

$$x'_n = x'_0 + n\frac{\delta B}{B\rho}\Delta s, \tag{7.10}$$

for the *nth* turn. We also know that $x_{n\,\text{MAX}} = \sqrt{\mathcal{C}\beta_n}$, $x'_{n\,\text{MAX}} = \sqrt{\mathcal{C}\gamma_n}$, where β, γ are the Courant–Snyder parameters for the horizontal motion, and $\mathcal{C}$ is the Courant–Snyder invariant (chapter 3). Therefore,

$$x'_{n\,\text{MAX}} = x_{n\,\text{MAX}}\sqrt{\frac{\gamma_n}{\beta_n}} = x_{n\,\text{MAX}}\frac{\sqrt{1+\alpha_n^2}}{\beta_n}. \tag{7.11}$$

Finally, by setting $x'_n = x'_{n\,\text{MAX}}$, $x'_0 = 0$, we get

$$x_{n\,\text{MAX}} = n\frac{\delta B}{B\rho}\Delta s\frac{\beta_n}{\left(1+\alpha_n^2\right)^{1/2}} \rightarrow n\frac{\delta B}{B\rho}\Delta s\beta_{n\,\text{max}}. \tag{7.12}$$

In the last step, we have used the fact that for $\alpha_n = 0$ we have $\beta_n = \beta_{n\,\text{max}}$, i.e., we have assumed that the kick error occurs at the location of maximum β. This is exactly the case for kicks caused by quadrupole displacement errors, since the quads are placed at the maxima (or minima) of the betatron function, but it is only an approximation for an arbitrary location of a bending dipole with an error. If there are N_d dipoles in the ring, on the other hand, with a field error distribution that has an *rms integrated error* of $(\delta B/B)\Delta s$, equation (7.12) has to be multiplied by $\sqrt{N_d}$ to obtain the total orbital excursion. Figure 7.6 illustrates the change in orbit due to an integer resonance. Additional considerations are given in [8].

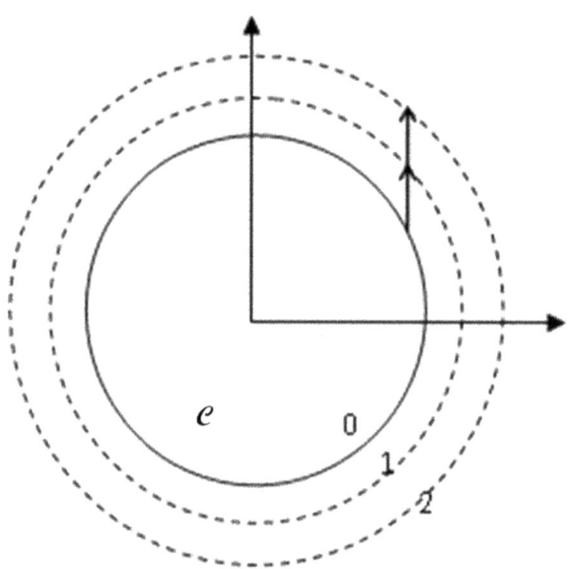

Figure 7.6. Schematic representation of an integer resonance—a bird's-eye view of how a particle is deflected by a dipole error kick into an orbit that grows linearly.

As an example, we consider the compact UMER (see references): $N_d = 36$, $\Delta s = 0.0376$ m, $\beta_{x\,max} = 0.54$ m (at $\nu_{0X} = 6.0$), $\rho = 0.275$ m. Therefore, for a 1% rms dipole field error $[(\delta B/B) = 0.01]$, we get an orbital change of 4.4 mm per turn; the beam centroid would reach the vacuum pipe in less than 6 turns. Integer resonances are very destructive! We have implemented a simulation of the resonance in the *WINAGILE* code (see appendix A); we include further details and hyperlinks to movies at the end of the chapter.

We can base a simple general treatment of resonances on the first equation of motion in equation (3.18), which we copy for reference:

$$x''(s) + \kappa_x(s)x = 0. \tag{7.13}$$

We now apply *Floquet's transformation* of variables to equation (7.13),

$$\phi(s) = \frac{\psi(s)}{\nu_{x0}}, \quad \eta(s) = \frac{x(s)}{\sqrt{\beta(s)}}, \tag{7.14}$$

to obtain

$$\frac{d^2\eta(\phi)}{d\phi^2} + \nu_{x0}{}^2\eta(\phi) = 0. \tag{7.15}$$

Equation (7.15) is a simple harmonic oscillator equation for the normalized variable η, and ν_{x0} can be recognized as the bare tune (no SC included), introduced in chapter 3.

With field errors $\Delta B(\eta, \phi)$ representing all fields not defining the reference orbit, i.e., all random field errors due to magnet construction, we now have the inhomogeneous equation [9, 10]:

$$\frac{d^2\eta(\phi)}{d\phi^2} + \nu_{x0}^2\eta(\phi) = -\nu_{x0}^2\beta^{3/2}\frac{\Delta B(\eta, \phi)}{(B\rho)}, \tag{7.16}$$

where $(B\rho)$ is the magnetic rigidity, and $\Delta B = B_0[b_0 + b_1(\phi)x + b_2(\phi)x^2 + ...]$ is a 1D (not the most general) expansion of the field errors. Therefore, equation (7.16) can be written as

$$\frac{d^2\eta(\phi)}{d\phi^2} + \nu_{x0}^2\eta(\phi) = -\frac{\nu_{x0}^2 B_0}{(B\rho)}[\beta^{3/2}b_0 + (\beta^{4/2}b_1)\eta + (\beta^{5/2}b_2)\eta^2 + ...]$$

$$= -\frac{\nu_{x0}^2 B_0}{(B\rho)}\sum_{p=\text{integer}} C_k e^{\pm ip\phi}, \tag{7.17}$$

where we have written the inhomogeneous part of the equation as a sum of *driving terms*. An *integer resonance* is excited if the dipole error represented by $\beta^{3/2}b_0$ is equal to $C_p e^{\pm ip\phi}$, with $p = \nu_{x0} = $ integer; a *half-integer resonance* occurs if $\beta^2 b_1 = C_p e^{\pm ip\phi}$ such that $p - \nu_{x0} = \nu_{x0}$, i.e., $p = 2\nu_{x0}$. The difference $p - \nu_{x0}$ arises from the beat frequency between the pth harmonic of the quadrupole gradient error $\beta^2 b_1$ and the oscillation at frequency ν_{x0} contained in the factor η in equation (7.17). By similar reasoning, the first nonlinear resonance, or *third-integer resonance*, occurs if $p - 2\nu_{x0} = \nu_{x0}$, or $p = 3\nu_{x0}$.

We assumed a superperiodicity $N = 1$ in the Fourier expansion of equation (7.17), second equality; in other words, the period of expansion was *one entire turn* in the circular accelerator. If the lattice has higher periodicity, however, the period of expansion is equal to the length of the shorter repeating structure. Thus, in an ideal 'ring' with six sections or superperiods, we would have $N = 6$, with a correspondingly shorter period in the Fourier expansion. In an actual ring, the superperiodicity may be broken by the injection line, which may contain special magnets and other elements. We can graphically represent the resonance condition equation (7.7) by plotting ν_{y0} vs. ν_{y0} for a range of values of n_x, n_y, and p. The *resonance diagram* with lines up to the third order, and with $N = 1$, is shown in figure 7.7.

If the superperiodicity is $N > 1$, i.e., higher symmetry, the resonance condition, equation (7.7), leads to more widely spaced resonances, which is desirable. However, these resonances are generally strong. *Sum resonances* occur when $\nu_{x0} + \nu_{y0} =$ integer, while *difference resonances* are characterized by $\nu_{x0} - \nu_{y0} =$ integer. These resonances are examples of *coupling resonances* arising from coupled motion in the transverse (betatron) degrees of freedom. The coupling may arise from

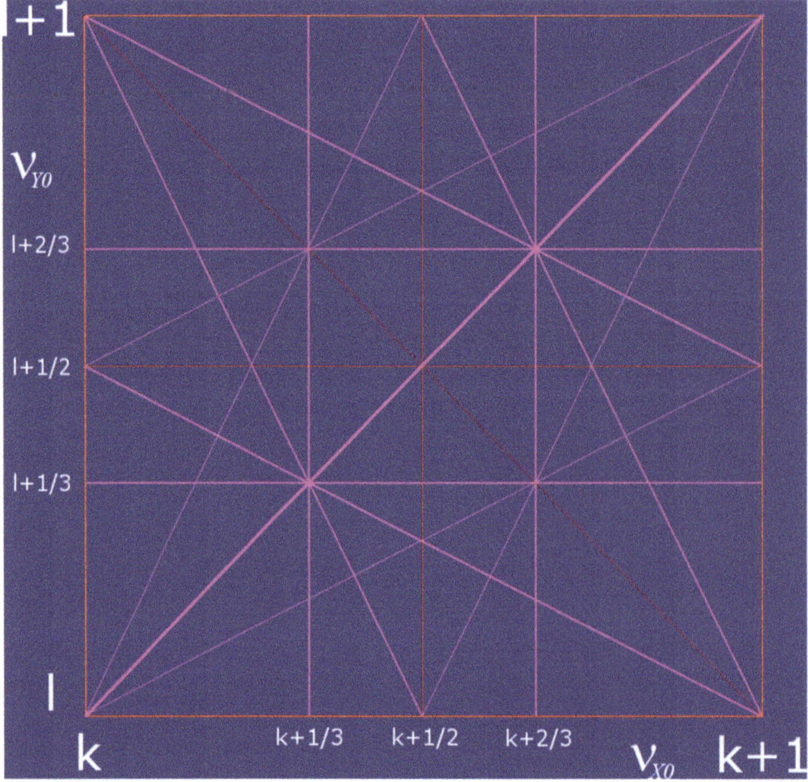

Figure 7.7. Resonance diagram for first-, second-, and third-order resonances and superperiodicity $N = 1$. The numbers k and l are integers. The diagram is based on an original WINAGILE chart. See also [11].

quadrupoles rotated due to misalignment, from skew quadrupole components present even in leveled magnets, or from solenoid fields. It turns out that the difference resonances are *stable*, while the sum resonances are *unstable* (see e.g., [10], chapter 5).

The actual resonance lines have width, unlike the chart in figure 7.7. Thus, the resonances are often called *stop bands*, and their widths are narrower the higher the order of the resonance. Furthermore, radiation damping reduces the growing betatron amplitude near a resonance.

A more general treatment of resonance theory can be found in a 1978 monograph by G Guignard [12].

7.4 Betatron resonances and space charge

The resonance condition expressed in equation (7.7) must be modified when direct SC is significant:

$$n_x(\nu_{x0} - \Delta\nu_{X\,\mathrm{eff}}) \pm n_y(\nu_{y0} - \Delta\nu_{Y\,\mathrm{eff}}) = p, \tag{7.18}$$

where $\Delta\nu_{X,\,Y\mathrm{eff}}$ are *effective* tune shifts, and we have assumed $N = 1$. To simplify further, we write $n(\nu_0 - \Delta\nu_{\mathrm{eff}}) = p$ for an isotropic beam. Following the discussion in section 7.3, 'p' represents the harmonic number of a field perturbation. The effective tune shift is related to the direct SC tune shift $\Delta\nu = \nu_0 - \nu$ through a coefficient C:

$$\Delta\nu_{\mathrm{eff}} = C\Delta\nu. \tag{7.19}$$

The coefficient C, introduced by Baartman [6] and inspired by work from Hofmann [4, 5], is called a 'coherent mode coefficient.' It can be interpreted as an attenuation coefficient of the single-particle SC tune shift by the collective (or 'coherent') motion if $C < 1$. This attenuation, which allows greater direct SC tune shifts before resonance occurs, is termed 'coherence advantage' by Hofmann [5].

In a simple-minded picture, $C = 1$, meaning that the resonance condition is shifted by just the SC incoherent tune change. However, as discussed in [6], the beam SC collective *modes* and not the individual particles resonate with field errors. Thus, the relevant tune is the *mode tune*, i.e., the number of mode oscillations per turn in a ring lattice, or, in terms of phase advance, the number of mode oscillations per lattice period multiplied by 2π. For isotropic beams, equation (7.18) is written as

$$n(\nu_0 - C\Delta\nu) = p. \tag{7.20}$$

The coherent mode coefficient in the simple form introduced in equation (7.19) is actually applicable to linear resonances only, i.e., $n = 1, 2$. For $n = 1$, corresponding to a dipole resonance, $C = 0$: the condition for integer resonances is not changed by SC [6]. When $n = 2$ in equation (7.20), on the other hand, we have a quadrupole, i.e., half-integer resonance. In this case, $C = \frac{1}{2}, \frac{3}{4}$, corresponding to the symmetric and antisymmetric envelope modes, respectively, discussed in section 7.1 in the context of a mismatched K–V beam.

To examine the half-integer resonance in some detail, let us consider equation (7.4) in terms of tunes:

$$\nu_{\text{sym.}} = \sqrt{2\nu_0^2 + 2\nu^2}, \text{ symmetric mode,}$$
$$\nu_{\text{asym.}} = \sqrt{\nu_0^2 + 3\nu^2}, \quad \text{antisymmetric mode.} \tag{7.21}$$

With $n = 2$, $C = 1/2$ in equation (7.20) we have $\nu_0 - (\Delta\nu/2) = p/2$. We reproduce this last equation if we set $\nu_{\text{sym}} = p$, and neglect $\Delta\nu/\nu_0$ in the first equation of (7.21), i.e., mode resonance occurs if the harmonic 'p' coincides with the symmetric mode tune ν_{sym}. Likewise, setting $n = 2$, $C = 3/4$ in equation (7.20) we obtain $\nu_0 - (3/4)\Delta\nu = p/2$, which can be reproduced with the second equation (7.21) if we set $\nu_{\text{asym}} = p$ and neglect $\Delta\nu/\nu_0$.

Using the linearized Vlasov equation and a water-bag particle distribution in a 1D model with an initial SC perturbation, Okamoto and Yokoya have shown [13] that equation (7.20) should be modified to read

$$n(\nu_0 - C\Delta\nu) = \frac{p}{2}. \tag{7.22}$$

The previous discussion of the half-integer ($n = 2$) resonance implies that the mode tunes can take *half-integer* values as well. In fact, the extra factor of ½ on the right-hand side of equation (7.22) is necessary to account for the envelope instability or 'parametric resonance' briefly discussed at the end of section 7.1. In this case, the phase advances per period of the two envelope modes are equal to 180°, indicating that the modes undergo half an oscillation per lattice period [2]. Herein some confusion exists between 'resonance' and 'instability': while a resonance phenomenon requires a *periodic* driving force, instability arises from just a *perturbation*. Thus, equation (7.20) is a condition for a mode resonance driven by external focusing through the harmonic 'p,' while equation (7.22) includes the additional case of a perturbation caused by the SC potential.

The expressions in (7.21) are valid for the case of equal tunes in the transverse directions. When the bare tune split is significant, i.e., of the order of the incoherent tune shift [6], equation (7.6) should be used instead. For a beam that is 'not too flat,' i.e., $\eta_0 \simeq 1$, the equation can be recast as

$$\nu_{\text{asym.}} = 2\left(\nu_{0x} - \frac{5}{8}\Delta\nu_x\right), \tag{7.23}$$

with a similar equation for the y-direction. Therefore, the coherent mode coefficient is $C = 5/8$ in the large tune split case. As discussed in [6], the envelope oscillations are coupled in the small tune split case (equation (7.21)) but uncoupled otherwise. In chapter 9, we present additional considerations related to the role of SC on resonances in the small UMER and in linear Paul trap simulators.

Additional 'odd' second-order SC modes are discussed by Hofmann [5]. These arise from a skew SC distribution but may not be directly observable because of overlap with other, higher-order resonances. To include the latter as well as beam anisotropy, the coherent mode coefficient is written as C_{mk}, with m, k denoting azimuthal and radial indices [6]. Equation (7.22) is then changed to

$$m(\nu_0 - C_{mk}\Delta\nu) = \frac{p}{2}. \tag{7.24}$$

In this new notation we have $C_{11} = 0$, $C_{22} = \frac{1}{2}$, $\frac{3}{4}$. Hofmann also considers third- and fourth-order modes, which have four and five frequency solutions, respectively. For example, the highest-frequency third-order mode has $C_{33} = \frac{3}{4}$ for the equal-tunes case, and 19/24 for split tunes. (Note that in Hofmann's notation, the highest-frequency solutions are written $\omega_k = k(\nu_x + F_k\Delta\nu_x)$ for modes of order k (which is different from index 'k' in equation (7.24)) and in terms of incoherent tune ν; our C coefficient is then $C = 1 - F$.) As pointed out by Baartman and others [6], some of the higher-order envelope modes of the theory reflect the nature of the ideal K–V distribution more than realistic distributions and thus are not expected to be observed. Other (initial) particle distributions, more notably the water-bag and Gaussian distributions, have been implemented in computer simulations and found to be in fair agreement with experimental results in a few large machines [5] (see also section 9.5).

7.5 Dispersion and space charge

In circular machines, the combined effect of dispersion and SC is to cause enlargement of the beam size in the bending (horizontal) plane beyond the effect of SC alone. In an alternating-gradient lattice, however, horizontal dispersion has an effect on the vertical plane because of coupling, although the effect is normally small. Furthermore, because of the reduced effective focusing caused by SC, for a given momentum error the main effect is an increase in dispersion. We can see this by considering equation (3.38) for the zero-current average linear dispersion: $\overline{D}_{0x} = \rho_m/\nu_0^2$. With SC, we could write $\overline{D}_x = \rho_m/\nu^2 = \overline{D}_{0x}/\eta^2$ as an ansatz, where η is the tune depression; thus, the average linear dispersion should increase because of SC. However, as we show in this section, the last expression overestimates the effect, although a quantity related to η should be used.

Equation (3.36) defines the effect of linear dispersion without SC. With SC we can write

$$x_{sc}s) = x_{sc\beta}(s) + D_x(s)\delta, \tag{7.25}$$

with the understanding that $D > D_0$ because SC simply enhances the displacements, and D still represents the closed orbit of off-momentum particles. However, as discussed in [14], measuring orbits with low and high current does not permit the measurement of D. Further, the reference orbit is not affected by incoherent SC, as the former is determined by the motion of the beam's centroid. Image forces, though, would affect the reference orbit for an off-center beam.

To measure the effects of incoherent SC on dispersion we need to evaluate the *second moments* of the beam distribution (and energy spread). Without SC we have

$$2x_{rms} = \left[a_0^2 + 4D_{0x}^2\langle\delta^2\rangle \right]^{1/2} = a_0\left(1 + \xi_0^2\right)^{1/2}, \tag{7.26}$$

where $\xi_0 \equiv 2D_0\Delta/a_0$, $\quad \xi_0^2 \ll 1$, with $\Delta = \langle\delta^2\rangle^{1/2}$ denoting the rms fractional momentum error or spread. (We recall from chapter 4 that $a_0 = \sqrt{\tilde{\varepsilon}_x/k_0}$.) Interestingly, the ratio D_{0x}/a_0 is fairly constant over a broad range of beam parameters and machines.

In an attempt to include SC, we could write, with $a \equiv 2(x_{\rm rms})_{\rm sc}$,

$$a = \left[a_{\rm sc}^{\,2} + 4D_x^2\langle\delta^2\rangle \right]^{1/2}, \tag{7.27}$$

which turns out to be approximately correct for small δ and with proper expressions for $a_{\rm sc}$ and D. The correct general expression that replaces equation (7.27) when *linear* SC and *linear* dispersion are combined is discussed next.

Regardless of SC, the standard horizontal *rms* emittance as defined in equation (4.7) is not conserved in the presence of dispersion. However, a generalized *conserved* emittance can be defined [15] as

$$\varepsilon_{dx}^2 = \varepsilon_{X\rm rms}^2 - \Delta^2\langle[x'D_x - xD'_x]^2\rangle. \tag{7.28}$$

Further, Venturini and Reiser have derived a set of equations for the envelope and dispersion functions of a continuous beam in the presence of linear SC and linear dispersion [16]:

$$X'' + k_{x0}^2X - \frac{2K}{X+Y} - \frac{\varepsilon_{dx}^2 + [XX' - D_xD'_x\Delta^2]^2}{X^3\left(1 - \dfrac{4\Delta^2 D_x^2}{X^2}\right)} + \frac{X'^2}{X}$$

$$- \frac{4\Delta^2}{X}\left(\frac{D_x}{\rho} + D'^2_x\right) = 0, \tag{7.29}$$

$$Y'' + k_{y0}^2X - \frac{2K}{X+Y} - \frac{\varepsilon_y^2}{Y^3} = 0,$$

where $\rho = \rho(s)$ is the local bending radius at the nominal energy. The dispersion function now obeys an equation that includes a SC term and is coupled to the transverse envelope 2rms semi-axis dimensions:

$$D''_x + \left[k_{x0}^2 - \frac{2K}{X(X+Y)}\right]D_x = \frac{1}{\rho}. \tag{7.30}$$

Compare this with equations (3.37) and (4.46). The weak focusing term $1/\rho^2(s)$ has been neglected in equation (7.30). Additional important general considerations regarding equations (7.29) and (7.30) can be found in [14].

By setting all derivatives in equations (7.29) and (7.30) to zero, we get a set of algebraic equations for the 'average' quantities in the uniform-focusing/bending approximation, just as was done in equations (4.46) and (4.47). Then, by solving the equation for a^2 we obtain, after some manipulation [14],

$$a^2 = \frac{a_0^2}{\eta_{x\Delta}} + 4\Delta^2 D_x^2, \quad \text{with} \quad \eta_{x\Delta} \equiv \sqrt{1 - \frac{2K}{a(a+b)k_{x0}^2}}, \tag{7.31}$$

and $D_x \equiv D_x(a, b, \Delta)$. The equation for a^2 has the form of equation (7.27); $\eta_{x\Delta}$ reduces to the standard tune depression η in the absence of dispersion ($\Delta = 0$). Thus, we have $a^2 = a_0^2/\eta$ in the limit of zero dispersion, and $a^2 = a_0^2(1 + \xi_0^2)$ in the limit of zero current (equation (7.26)). Further, the average horizontal dispersion is, from equation (7.30),

$$\overline{D}_x = \frac{\overline{D}_{0x}}{\eta_{x\Delta}^2}, \quad \overline{D}_{0x} = \frac{1}{\rho_m k_{x0}^2}. \tag{7.32}$$

(Compare equation (7.38) with equation (3.38).) With SC, the average dispersion increases, but not by as much as implied using standard tune depression. Further, by assuming $a \cong b$, $\epsilon_x = \epsilon_y$, $k_{x0} = k_{y0} \equiv k_0$ and small Δ, we arrive at our main result:

$$\frac{a^2}{a_0^2} \cong \frac{1 - \eta^2}{2\eta} + \frac{\xi_0^2}{2\eta^2} + \sqrt{\left(\frac{1 - \eta^2}{2\eta} + \frac{\xi_0^2}{2\eta^2}\right)^2 + 1} \; . \tag{7.33}$$

Equation (7.33) is valid over a broad range of average lattice dispersion and beam currents if SC is not extreme (η of the order of 0.1); in the latter situation, if $\Delta \geqslant 1\%$ the equation can only provide a fair estimate of beam radius. Figure 7.8 illustrates the results of calculations employing equation (7.33) as well as the smooth-approximation equations (data points) for five UMER beams, a proton storage ring, and a hypothetical ring for heavy-ion fusion (HIF).

We emphasize that there are two major, not entirely independent, assumptions behind the approximations used so far: first, both rms envelope and dispersion matching are satisfied or nearly so; and second, the rms emittance change is

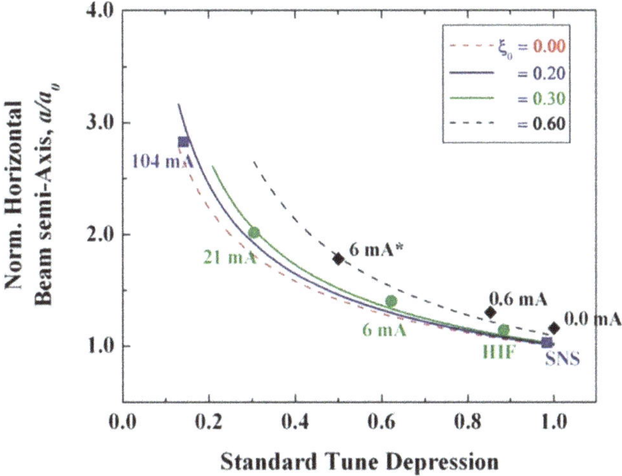

Figure 7.8. Combined effects of dispersion and direct SC: normalized horizontal beam semi-axis as a function of standard tune depression η, following equation (7.33). Four values of the parameter ξ_0 (see equation (7.26) and the text) are shown, with $\Delta = 0$ for the first case (red dashed line) and $\Delta = 0.01$ for the other three cases. The points represent exact smooth-approximation calculations for five cases of beams in UMER, the proton ring at the spallation neutron source in ORNL, and a hypothetical ring for heavy-ion fusion. See also [14].

negligible. Clearly, these assumptions may not be realistic for large momentum errors (>1%), for which case the smooth focusing/bending approximation may not be accurate, but those cases are considered extreme. Furthermore, equation (7.33) and figure 7.8 may give the impression that ξ_0 and η (standard horizontal tune depression) can be specified independently for a given beam current and emittance. This is not the case, as both ξ_0 and η are functions of $a_0 = (\varepsilon/k_0)^{1/2}$. To specify the beam transport problem completely in the smooth approximation and in the presence of both direct SC and dispersion, we need the current, energy, rms emittance, and rms fractional momentum error Δ (beam parameters), as well as the average machine radius and magnet strength (lattice parameters). The standard tune depression and the ratio D_0/a_0 are, in turn, independent of Δ, but ξ_0 is proportional to Δ. To illustrate this further, we have included in figure 7.8 a point labeled '6 mA*' on the curve for $\xi_0 = 0.60$; this point is possible if we adjust both the emittance and the momentum error ($\Delta = 0.015$) to yield $\xi_0 = 0.60$ and $\eta = 0.50$ while keeping the other parameters the same as for the standard transport of 6 mA in UMER [14]. In UMER, the effect of small Δ (0.2% or 40 eV uncorrelated energy spread) on the horizontal beam dimension is a 1% effect, independent of current; for large Δ (1% or 200 eV uncorrelated energy spread) the effect is 10%–20%, relative to direct SC effects alone. Experimental verification of the theory could involve the study of average beam dimensions as functions of rms uncorrelated energy spread, essentially equation (7.31), but there is no simple way to control the energy spread.

7.6 Computer resources

The computer codes MENV and SPOT for rms envelope matching including SC, as well as the specification files for the examples presented in section 7.1, are available on this book's website https://ireap.umd.edu/clark/faculty/1273/Santiago-Bernal. Appendix A contains a brief description of these codes. In addition, TRACE 2-D and TRACE 3-D, freely available from Los Alamos (see appendix A for hyperlink), are also very useful. The calculations that lead to figure 7.4 on mismatch envelope modes are contained in the Mathcad program **ENVELOPE_MODES-6mA-OneTurn.xmcd**. Calculations related to integer resonances (section 7.2) are presented in the Mathcad worksheet **IntegerResonance.xmcd**. Two movies, **IntRes_Movie-I.avi** and **IntRes_Movie-II.avi,** demonstrate the destructive effects of dipole and quadrupole errors in UMER.

Problem 10.2.3 in [17] is useful for understanding nonlinear resonances without SC. The problem is stated by adding a driving term to the basic equation (7.15) (in normalized variables):

$$\frac{d^2\eta(\phi)}{d\phi^2} + \nu_{x0}{}^2\eta(\phi) = \varepsilon\eta^p \cos(m\phi), \tag{7.34}$$

where ε, p, and m are constants representing the strength of the perturbation, its order ($p+1$), and the azimuthal harmonic, respectively. The problem is approached in [17] via a perturbation approach and also by considering a simple matrix calculation. The following matrix equation represents the result of iteratively applying a zero-length nonlinear kick for a specified number of steps N_{step}:

$$\begin{pmatrix} \eta_{j+1} \\ \eta'_{j+1} \end{pmatrix} = \begin{pmatrix} \cos \mu_{\text{step}} & \sin \mu_{\text{step}} \\ -\sin \mu_{\text{step}} & \cos \mu_{\text{step}} \end{pmatrix} \left[\begin{pmatrix} \eta_j \\ \eta'_j \end{pmatrix} + \begin{pmatrix} 0 \\ \varepsilon \cos\left(\dfrac{2\pi m}{N_{\text{step}}}j\right)\eta_j^p \end{pmatrix} \right], \tag{7.35}$$

$$\mu_{\text{step}} = \frac{2\pi\nu_0}{N_{\text{step}}}.$$

We use a program written in C to track a number of particles over a specified number of turns. The initial conditions of the succeeding particles vary according to a predefined rate, and the trace-space data is generated for all particles over the number of turns. The program as well as the executable file are available on this book's website. (The program was compiled in Ubuntu Linux 16.04, but the executable will also run in the Ubuntu app for Windows 10.) Figure 7.9(a) displays the trace space for 250 particles over 2048 turns for a third-order ($p = 2$) perturbation of unit strength; other parameters are displayed in the table inset. There are 2048 × 250 = 512 000 data points in (η, η') space, i.e., one 2048-point trace per particle.

Note that the small ellipses near the origin gradually distort for larger amplitudes and that four resonance islands appear inside a relatively wide gap between the 14th and 16th traces (i.e., the 13th and 14th particles). More islands can be seen near the edges of the trace-space region shown. Figure 7.9(b) shows the results of calculating the FFT for 4 traces in figure 7.9(a); the tune increases for larger amplitudes, and the four resonance islands correspond to a tune 0.250, i.e., a quarter-integer resonance. Thus, a third-order perturbation can lead to resonances higher than third-integer resonances. The discussion in [17] contains additional analytical insights.

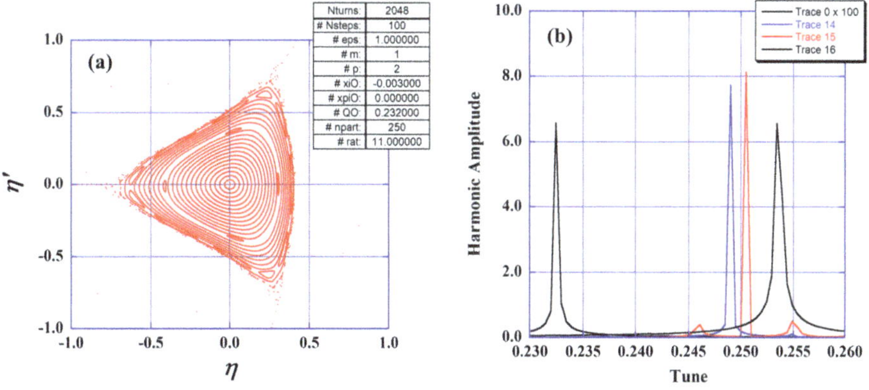

Figure 7.9. (a) Poincaré plot of 250 particles over 2,048 turns acted upon by a third-order perturbation ($p = 2$ in equation (7.35)). The table inset contains all the parameters of the calculation using the program NonLinRes. (b) Amplitude of FFT of traces 0, 14, 15, and 16; the four resonance islands (trace 15) correspond to a tune equal to 0.250. See also [17].

References

[1] Bryant P J and Johnsen K 1993 *The Principles of Circular Accelerators and Storage Rings* (Cambridge: Cambridge University Press)

[2] Reiser M 2008 *Theory and Design of Charged Particle Beams* 2nd edn (Weinheim: Wiley)

[3] Hofmann I, Laslett L J, Smith L and Haber I 1983 *Part. Accel.* **13** 145

[4] Hofmann I 1998 Stability of Anisotropic Beams with Space Charge *Phys. Rev. E* **57** 4713

[5] Hofmann I 2017 *Space Charge Physics for Particle Accelerators* (Berlin: Springer)

[6] Baartman R 1998 *Betatron Resonances with Space Charge Conf. Proc. No. 448* (New York: AIP)

[7] Bernal S, Li H, Kishek R A, Quinn B, Walter M, Reiser M and O'Shea P G 2006 RMS envelope matching of electron beams from 'zero' current to extreme space charge in a fixed lattice of short magnets, *Phys. Rev. ST Accel. Beams* **9** 064202

[8] Le Duff J 1979 Integer Resonance Crossing in H.I. Accumulator Ring, *Proc. of the Heavy-Ion Fusion Workshop (Berkeley, CA, 29th October—9th November)* p 310

[9] Wille K 2000 *The Physics of Particle Accelerators*, An Introduction (Oxford: Oxford University Press), Sec. 3.14.3

[10] Edwards D A and Syphers M J 2004 *An Introduction to the Physics of High Energy Accelerators* (New York: Wiley-VCH) ch 4, 5

[11] Wiedemann H 2007 *Particle Accelerator Physics* 3rd edn (Berlin Heidelberg: Springer) ch 13

[12] Guignard G 1978 *A General Treatment of Resonances in Accelerators, CERN report CERN 79-11*

[13] Okamoto H and Yokoya K 2002 Parametric resonances in intense one-dimensional beams propagating through a periodic focusing channel *Nucl. Instrum. Methods Phys. Res A* **482** 51–64

[14] Bernal S, Beaudoin B, Koeth T and O'Shea P G 2011 Smooth approximation model of dispersion with strong space charge for continuous beams *Phys. Rev. ST Accel. Beams* **14** 104202

[15] Venturini M, Kishek R A and Reiser M 1999 *Proc. 18th Particle Accelerator Conf. (New York)*

[16] Venturini M and Reiser M 1998 *Phys. Rev. Lett.* **81** 96

[17] MacKay W W and Conte M 2012 *Accelerator Physics: Example Problems with Solutions* (Singapore: World Scientific)

IOP Publishing

A Practical Introduction to Beam Physics and Particle Accelerators (Third Edition)

Santiago Bernal

Chapter 8

Linacs and rings (examples), closed orbit, and beam cooling

In the first half of this chapter, we show how to employ theory developed in the previous chapters to verify typical parameter tables found on the home websites of accelerators. Thus, in section 8.1 we give examples of two *linacs* and their characterization, and in section 8.2 we illustrate calculations for two ring light sources. (Additional examples of research ring accelerators and storage rings are given in chapter 9.) In section 8.3, we discuss the important concept of the *closed orbit* in ring accelerators, perhaps better characterized as the *equilibrium orbit;* this section is an extension of the material on dispersion and resonances provided in the previous chapters. We conclude section 8.3 with a brief description of the orbit response matrix and corrections. Lastly, in section 8.4 we give a succinct account of the techniques of beam cooling, with some details on stochastic cooling.

The 'Computer Resources' section includes Mathcad worksheets as well as MAD-8 and WINAGILE files for the machines discussed.

8.1 Examples of linacs

In this section, we present two examples of linear accelerators: the SLAC electron linac, a facility near Stanford University in California, and the Low-Energy Demonstration Accelerator (LEDA) proton linac at Los Alamos National Laboratory.

The original SLAC linac accelerated electrons to 50 GeV over a distance of about 3 km. The basic 3 m accelerating unit of the SLAC linac has become an industry standard and is in use in many normal conducting linacs around the world. The unit, briefly mentioned in chapter 6, is an S-band, disk-loaded, constant-gradient, traveling-wave structure. Table 8.1 summarizes its main parameters.

The first calculation from the data in the table is the RF wavelength and its relationship to the iris separation d: $\lambda_{RF} = 0.105$ m, $\lambda_{RF}/d = 3$. This last result reflects the $2\pi/3$ operating mode of the cavity (see figure 6.1), or, equivalently, the 120° phase advance between cells. The number of cavities is 86, which is close to L/d. Further, and as discussed in chapter 6, the tapered construction of the cavity geometry (a gradual reduction in $2a$ and $2b$) guarantees that the peak accelerating field is constant, i.e., that the structure has a constant gradient. The variation of the group velocity along the 3 m structure can be calculated using equation (6.7): $v_g(0) = 0.018c$, and $v_g(L) = 0.006c$. The *filling time* is $t_F = 2\tau Q/\omega_{RF} \approx 1$ µs.

The energy gradient can be calculated using equation (6.8); we obtain 62.0 MeV over 3.05 m, i.e., 20.3 MeV/m. The RF peak power of the original design, 8 MW, leads to a gradient equal to 9.73 MeV/m. Finally, the energy at the end of a 3 km linac comprising such 3 m sections would be equal to 61 GeV. The actual linac uses 960, 3 m structures.

The second example of a linac is a proton machine at Los Alamos. Table 8.2 contains the parameters of the LEDA proton linac collected from publications related to experiments on beam halo conducted around 2001 (see [1]). The proton beam is initially accelerated and focused by an 8 m, 350 MHz RF quadrupole

Table 8.1. Main parameters of the SLAC 3 m linac unit.

RF frequency, f_{RF}	2856 MHz
Shunt impedance, r_s	53–60 MΩ/m
Length, L	3.048 m
Iris separation, d	35.001 mm
Disk diameter, $2b$	8.4–8.2 cm
Iris diameter, $2a$	2.6–1.9 cm
Number of cells	86
Quality factor, Q	13000–14000
Phase between cells	120°
Accelerating gradient	>17 MV m^{-1}
Attenuation constant, τ	0.57
RF filling time, t_F	0.95 µs
Peak RF power, P_{r0}	35 MW

Table 8.2. Main parameters of the LEDA halo experiment at Los Alamos [1].

Energy, E	6.7 MeV
Particle	Proton
Total length, L	11 m
Lattice period, S	0.42 m
Beam current continuous wave, I_b	<75 mA, CW
Emittance (root-mean-square (rms), unnorm.), $\tilde{\varepsilon}_{un.}$	>3.0 µm
Undep. phase advance, σ_0	80°
Phase adv. depression, $\sigma/\sigma_0 = \eta$	0.82–0.95
Beam radius (rms matched), a	>1.1 mm

(RFQ—see, for example, [2]) and matched with the first four quadrupoles into an 11 m strong-focusing lattice. Several beam diagnostics allow measurements of the beam profiles and emittances under different matching conditions.

Envelope mismatch is a major factor that contributes to halo formation. As seen at the end of section 7.1, the envelope of a mismatched beam undergoes symmetric ('breathing') and antisymmetric ('quadrupole') envelope modes. These modes can resonate with particles in the core of the beam, expelling the particles to a maximum radius that depends on the degree of mismatch and the size of the core. We use data in the table to verify the quoted 'tune' depression η, effective beam radius with and without space charge (a, a_0), and the number of envelope mode oscillations over the length of the beamline. We find $\eta = \sigma/\sigma_0 = 0.97$ and (a_0, a) = (1.90, 1.93) mm using the equations in section 4.4. Thus, beam transport in the halo experiments at LEDA was emittance dominated, but space-charge effects were appreciable, much more so than in electron linacs after acceleration. The values for 'tune' depression in the table are smaller but not too different from our result. Further, the effective matched beam radii (see the Kapchinskij–Vladimirskij (K–V) distribution in chapter 4) must be divided by two to obtain the rms value; thus we have (a_{rms}, $a_{0\mathrm{rms}}$) = (0.95, 0.97) mm, only about 10% smaller than the value in table 8.2.

To compute the number of envelope oscillations over 11 m, we first find the phase advances per period for the symmetric and antisymmetric modes. Using equations (7.4) and a 0.97 'tune' depression, we obtain (σ_{sym}, σ_{asym}) = (157°, 156°); from the wave numbers $k_{\mathrm{sym}} = \sigma_{\mathrm{sym}}/S$ and $k_{\mathrm{asym}} = \sigma_{\mathrm{asym}}/S$ we get $L/\lambda_{\mathrm{sym}} = 11.4$ oscillations, and $L/\lambda_{\mathrm{asym}} = 11.3$ oscillations. The original paper (see [1]) reports 'about ten mismatch oscillations.'

In conclusion, the K–V distribution and smooth-approximation models yield good results even when applied to emittance-dominated beam transport problems, as in the LEDA experiments.

8.2 Examples of rings

We present here two examples of synchrotron radiation (SR) sources (chapter 6): the first-generation (ca. 1970), *weak-focusing* machine, the Synchrotron Ultraviolet Radiation Facility II (SURF II) at NIST in Gaithersburg, MD in the U.S., which is not exactly a 'ring'; and one of the first second-generation light sources (1982), the double-bend achromat (DBA), the National Synchrotron Light Source (NSLS) vacuum ultraviolet (VUV) storage ring at Brookhaven National Laboratory (BNL) in Upton, NY. Table 8.3 summarizes the parameters of SURF II. The SURF II lattice is one of the examples in the code WINAGILE (see appendix A); a MAD input file and additional information can be found in Jim Murphy's Data Book [3]. Further, a working MAD-8 input file for SURF II (SURFII.MAD) can be found in this book's website. In these files, the single bending magnet of SURF II is modeled using eight 45°-sector magnets. The numbers for the tunes and chromaticity in table 8.3 can be extracted from the WINAGILE or MAD-8 files, but it is instructive to derive some of them from available relations. First, the *field index n* (chapter 2) can be deduced to be $n = 0.595$; this follows from equation (2.28) and the fact that

Table 8.3. SURF II parameters.

Energy, E	0.30 GeV
Superperiod, N	1
Bending radius, ρ	0.837 m
Betatron horizontal tune, ν_{0X}	0.640
Vertical betatron tune, ν_{0Y}	0.768
Momentum compaction, α	2.44
Emittance (natural), ε	0.350 μm
Natural hor. chromaticity, ξ_X	−1.79
Natural vert. chromaticity, ξ_Y	1.49
Transition gamma, γ_{tr}	0.640

the focusing constant ('K1' in MAD-8) $\kappa = 0.8496$ m^{-2} applies to the vertical plane. The betatron tunes are then obtained from equations (2.30) and (2.31); naturally, for a weakly focusing machine, the betatron tunes are less than 1.0. The momentum compaction factor and transition gamma are found using equations (3.42) and (3.44). The betatron function and dispersion values (not shown in the table) as well as the chromaticity can also be found using the equations in chapter 3. Finally, the *natural emittance* in μm can be obtained from equation (6.26) after adapting the result to a weakly focusing machine (see [3], p 41):

$$\varepsilon\,[\mu m] = \frac{C_q \gamma^2}{n\sqrt{1-n}}, \quad C_q = 3.83 \times 10^{-13}\ \mathrm{m}. \tag{8.1}$$

Important additional calculations pertain to the SR that can be produced with SURF II. First of all, note from table 8.3 and equation (6.13) the scaling of the radiated power per electron; SURF II has a very small radius but also low energy relative to other SR sources. Furthermore, the synchrotron radiation obtained from SURF II has a continuous spectrum with a *critical energy* (chapter 6) given by equation (6.28). We obtain $E_{\mathrm{cr}} = 0.072$ keV, or a critical wavelength ($h =$ Planck's constant) of $\lambda_{\mathrm{cr}} = hc/E_{\mathrm{cr}} = 17$ nm, which is outside the 'water window' between 2.34 and 4.4 nm. The total radiated power per electron is, according to equation (6.13), $P_R = 7.9$ nW. Assuming a current of $I_B = 100$ mA, we can calculate the number of electrons to be $N_e = 2\pi\rho I_B/e\beta c = 1.1 \times 10^{10}$. Therefore, the total irradiated power is $P_T = N_e P_R = 86$ W. A good exercise for the reader is to repeat the calculations for SURF III, which has a higher energy (0.38 GeV) and an adjustable field index n. The SURF III website http://physics.nist.gov/MajResFac/SURF/SURF/accjavan.html is a good guide. We now turn our attention to a strong-focusing ring.

Table 8.4 summarizes the parameters of the NSLS VUV, or NSLS-I. (The machine has been replaced by the third-generation 3.0 GeV NSLS-II light source, which started operation in 2015.) The basic cell of NSLS-I is a DBA or *Chasman–Green lattice*. It consists of two bending dipoles with a quadrupole doublet between them; sextupoles are also present to correct the chromatic aberration of the quadrupoles. See [4] for more details about the DBA. Figure 8.1 illustrates the

Table 8.4. Main parameters of the second-generation light source NSLS VUV.

Energy, E	0.808 GeV
Superperiod, N	4
Circumference	51.02 m
Horizontal betatron tune, ν_{0X}	3.123
Vertical betatron tune, ν_{0Y}	1.178
Synchrotron tune, ν_{0S}	0.0018
Momentum compaction, α	0.0235
Equilibrium hor. emittance, ε_x	0.154 μm
Natural hor. chromaticity, ξ_X	−3.44
Natural vert. chromaticity, ξ_Y	−5.77
Transition gamma, γ_{tr}	6.52

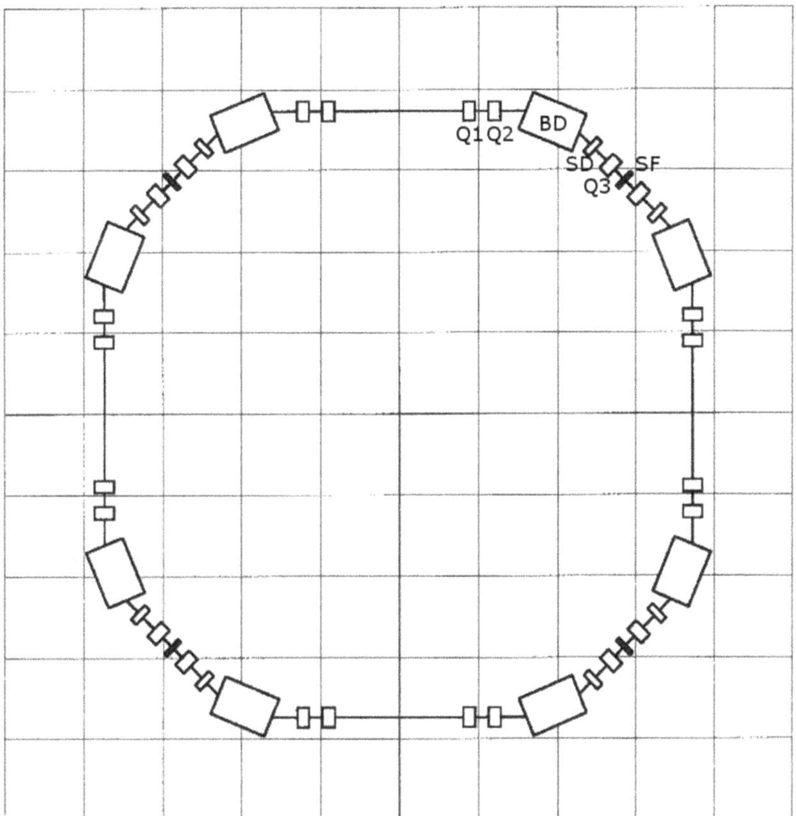

Figure 8.1. Lattice schematics of the second-generation light source NSLS VUV. The basic half-cell on the upper right is labeled as follows: 'Q' denotes a quadrupole, 'BD' a bending dipole, and 'SD, SF' are sextupoles. The grid size is 2.00 m. Adapted from WINAGILE code output.

basic lattice of NSLS VUV. The betatron tunes are greater than one, as expected for a strong-focusing machine. The numbers in table 8.4 are taken from calculations performed using WINAGILE and MAD-8 and from an old BNL website. We obtain the same tunes in WINAGILE and MAD-8, but the website quotes (3.14, 1.26); the chromaticities are obtained from MAD-8. Figure 8.2 shows the graphical output produced by MAD-8.

Additional calculations for the SR and damping are based on the SR integrals whose equations are presented in chapter 6. The results for these integrals are summarized in table 8.5; the values were extracted from the 'Synchrotron Radiation Data' feature in WINAGILE.

The default *total particle energy* in MAD-8 is 1.0 GeV, while WINAGILE uses 1.0 GeV as default for the *kinetic energy*. Naturally, the total and kinetic energies are very close to each other for highly relativistic particles, but they are not identical. Furthermore, the strengths of the magnets (e.g., 'K1' for quadrupoles in MAD-8, MAD-X, ELEGANT, and other accelerator codes) are independent of energy, as

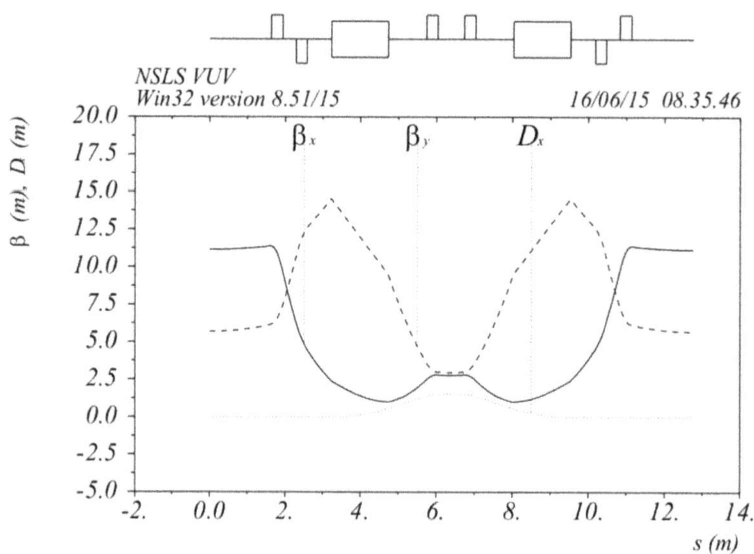

Figure 8.2. MAD-8 graphical output of the lattice functions for NSLS VUV. The magnet layout on top corresponds to a complete cell. The superperiodicity of the ring is $N = 4$.

Table 8.5. Synchrotron radiation integrals for NSLS VUV (from WINAGILE).

I_1	1.199775 m
I_2	3.289868 m^{-1}
I_3	1.722571 m^{-2}
I_4	−0.246102 m^{-1}
I_5	0.567642 m^{-1}

they are given as focusing constants. The radiation integrals, partition numbers, and momentum compaction are likewise independent of energy, but other derived quantities, such as energy loss per turn, energy spread, equilibrium emittance, and damping times depend very critically on the energy and the type of particle.

We use equation (6.17) to calculate the energy loss per turn; the total radiated power for a 1.0 A beam current at NSLS VUV is then 19.7 kW. Further, the *natural energy spread* and the *equilibrium horizontal emittance* are 5×10^{-4}, and 0.154 μm, respectively (see equation (6.26)). To find the damping times, we need the bending radius ρ of the dipoles. We get ρ from the parameter sheet at the old NSLS VUV website; the magnetic rigidity quoted there was $B\rho = 1.41$ Tesla $\times$ 1.91 m. Therefore, the damping times are (see equations in chapter 6) $\tau_x = \tau_y = 13$ ms and $\tau_\varepsilon = 7.2$ ms.

8.3 Closed orbit and correction

In its most obvious definition, the *closed orbit* (CO) in a ring lattice is the trajectory that closes in itself in position *and* slope, (x, y), and (x', y'), for the radial (x) and vertical (y) planes. More precisely, a CO is an *equilibrium orbit* satisfying Hamilton's variational principle: the magnetic flux linked by the orbit has a stationary value, which is minimized for stable motion. The mathematical details of this general concept are discussed in, for example, the last chapter of [5] and in appendix C of [6]. In this section, we discuss a few additional considerations for practical use, which are covered in more detail in [6–8].

The beam centroid displacement in the radial direction as measured by a beam-position monitor (BPM) at a given location s in a ring lattice includes contributions from the CO (also called 'closed orbit distortion'), the betatron function, dispersion, and BPM error:

$$x_{\text{BPM}}(s) = x_{c.o.}(s) + x_\beta(s) + x_\delta(s) + \varepsilon_{\text{BPM}}. \tag{8.2}$$

The second and third terms above comprise the displacement expressed in equation (3.36). We have $x_\delta = D_{0x}(s)\delta$ for linear dispersion and no space charge (δ is the fractional momentum error). The last term denotes a BPM error caused by calibration and other sources. With no magnet, momentum, or BPM errors, the BPM would directly measure the betatron oscillations whose amplitude is proportional to $\beta_x^{1/2}(s)$ (equation (3.19)). Because the betatron term is $x_\beta(s) = x_{\text{BPM}}(s) - x_{c.o.}(s)$, with magnet but no momentum errors, the CO defines the *reference (or equilibrium) orbit* for particles of design momentum p_0. Confusion may arise here because the 'reference orbit' is often understood to be the ideal orbit with design momentum p_0 and no magnet errors, but betatron oscillations *should not* be defined relative to this orbit.

The value of $x_{c.o.}(s)$ is easily calculated in the presence of a magnet error corresponding to a kick θ (in radians). We apply the one-turn matrix (equation (3.27) with $\Delta\psi = 2\pi\nu_x$) to an initial vector $(x_0, x_0')^T$ and add a kick represented by the vector $(0, \theta)^T$; we then impose the condition that the result should equal the initial vector, i.e., that the orbit should be closed. The math is greatly simplified by the use of the complex exponential form of the transfer matrix [9]; the result for the closed orbit is:

$$x_{c.o.}(s) = \frac{\beta_x^{1/2}(s)}{2 \sin \pi \nu_x} \beta_{x0}^{1/2} \theta \, \cos \left[\psi_x(s) - \pi \nu_x \right]. \tag{8.3}$$

The kick of θ may arise from a mispowered dipole magnet or a displaced quadrupole. A simplified version of this equation was derived in chapter 7 (see equation (7.12)) for an integer resonance (ν_x = integer) driven by the dipole kick $\theta = (\delta B/B_0 \rho) \Delta s$ caused by a field error $\delta B/B_0$ in a magnet of length Δs. As in chapter 7, the root-mean-square closed orbit displacement resulting from uncorrelated magnet errors around the ring lattice can be obtained by adding the square of terms of the form of equation (8.3) with θ replaced by θ_i for the ith magnet and taking the average around the ring [6].

If the error does not arise from from a magnet but rather from momentum, the radial displacement $x_\delta = D_{0x}(s)\delta$ can be found by solving equation (3.37). If the weak focusing term is ignored, which is a good approximation for large rings, the horizontal dispersion is found to be [6]

$$D_{0x}(s) = \frac{x_\delta(s)}{\delta} = \frac{\beta_x^{1/2}(s)}{2 \sin \pi \nu_x} \int_s^{s+C} \frac{\beta_x^{1/2}(\zeta)}{\rho(\zeta)} \cos \left[\psi_x(s) - \pi \nu_x \right] d\zeta. \tag{8.4}$$

Mathematically, the solution of the ordinary differential equation in equation (3.37) is found to be the sum of the solution of the homogeneous equation and a *particular solution*. The latter can be obtained by the method of variation of parameters, which involves integrals of the *cosine-like* and *sine-like* functions introduced in chapter 1 and the inhomogeneous term in equation (3.37). This is the method followed in reference [6], but a shortcut via the one-turn transfer matrix that led to equation (8.3) (see [9]) is equally valid.

The resemblance between equations (8.4) and (8.3) is clear. First, a momentum error 'kick' can be written as $\theta = \left(\delta p/p_0 \right)\rho^{-1}\Delta\zeta = (\delta/\rho)\Delta\zeta$, where ζ is a dummy variable representing arc length. Second, the momentum error is distributed continuously, so an integral encompassing the ring 'circumference' must be used. Third, the dispersion function D_{0x} can be interpreted as the closed orbit of off-momentum particles in the absence of magnet errors.

In practice, the CO is obtained by taking the average of thousands or more turns in a ring, but initial tuning may require the CO to be directly obtained from BPM data for four consecutive turns. Again, the one-turn transfer matrix, equation (3.27), can be used to relate the initial conditions $(x_{\beta 0}, x_{\beta 0}')^T$ to the final conditions after n turns, $(x_n, x_n')^T$, and after $n+1$ turns, $(x_{n+1}, x_{n+1}')^T$. It is found that the tune and the CO distortion are given by [6–8]:

$$\cos 2\pi \nu_x = \frac{x_n - x_{n+1} + x_{n+2} - x_{n+3}}{2(x_{n+1} - x_{n+2})}, \tag{8.5}$$

and

$$x_{c.o.} = \frac{x_{n+1}^2 - x_{n+2}^2 + x_{n+1}x_{n+3} - x_n x_{n+2}}{3(x_{n+1} - x_{n+2}) - x_n + x_{n+3}}. \tag{8.6}$$

Note that in equation (11.16) of [6] there is a factor of two in the denominator instead of three as in equation (8.6). However, in chapter 11 of [6], the use of the first two turns is not equivalent to *any* two consecutive turns as in equation (8.6). The latter has been explicitly derived by Sutter [8]. Furthermore, equation (8.5) provides just the fractional part of the tune. The integer + fractional (small-amplitude) tune can be measured, for example, by fitting a sinusoid to a difference of consecutive orbits.

An important component of tuning a ring accelerator is the measurement of the *orbital response matrix,* i.e., the change of orbital displacement x with respect to θ at each BPM. If the ith steering corrector magnet produces a kick θ_i, and the orbit change is observed at the jth BPM, then the orbit response matrix is

$$\boldsymbol{R}_{ij} = \frac{dx_j}{d\theta_i}, \tag{8.7}$$

which ideally yields sinusoidal terms modulated by the square root of the local beta function, as in equation (8.3) for the CO distortion. Clearly, the BPMs should be located where the beta functions have maximum values, i.e., at the center of the quadrupole magnets. In practice, the measured response matrix is fitted to a model response matrix to find the optics parameters that best represent the measurements. More systematically, the measured CO, as obtained from the four-turn formula (equation 8.6) is fitted to a model CO. In either case, the inversion of equation (8.7) with the best model response matrix should yield the corrector settings for the operating tunes and design energy of the machine. Matrix codes such as MAD, ELEGANT, and WINAGILE (see appendix A) can be used to compute COs and corrections via different approaches such as least-squares and single-value decomposition (SVD). Further, Linear Optics from Closed Orbits (LOCO) [10] is a more specialized code available for Matlab. At the end of chapter 9, we give an example of CO calculation and correction using WINAGILE.

8.4 Beam cooling

Beam cooling consists in the manipulation of a beam's particle distribution in order to reduce emittance and thus increase *brightness.* The latter is, in the transverse plane, $B \propto I_p/(\varepsilon_x \varepsilon_y)$, i.e., proportional to the phase-space density (I_p is the beam peak current). A good overview of beam cooling is given in [11, 12]; table 1 of [11] summarizes the main features of the different methods. The applicability of the latter depends on particle species (electrons vs. ions), machine configurations, energies, and beam current and dimensions. A beam can be cooled by another co-moving beam (e.g., the cooling of a heavy-ion beam by an electron beam); through processes such as ionization or laser cooling; by electronic-feedback methods (stochastic cooling); or by radiative cooling which involves damping motion via a combination of synchrotron radiation, random quantum photon emission, and RF acceleration.

Radiative or radiation cooling is used in all synchrotron light facilities and in damping rings for beam injection into electron linear colliders.

In *electron cooling*, suggested by Budker in 1966, heat flows from the hotter (ion) beam to the colder (electron) one in an irreversible way. At equilibrium, the two kinetic beam temperatures are equal, but the emittance of the ion beam is much smaller than that of the electron beam. The method cools all degrees of freedom simultaneously and works best for low-energy antiprotons and ions; typical cooling times are milliseconds to hours.

In *ionization cooling*, introduced by Skrinsky around 1980, energy is deposited into a material medium, but longitudinal momentum is recovered in RF cavities as in radiation damping (see chapter 6); this process is envisioned as a way to cool muon beams. Ionization cooling is limited by statistical fluctuations (straggling) and random multiple scattering.

In *laser cooling*, two counterpropagating laser beams are tuned to special closed-energy transitions of certain ions; cooling occurs in the longitudinal plane and is counteracted by heating via spontaneous photon emission. Laser cooling achieves the lowest ion beam temperatures, and its main long-term goal is the production of crystalline beams.

Common features of all cooling methods are the presence of external control agents (other beams, ionizing media, lasers, kicker-pickup feedback electronics, focusing/bending lattices, RF cavities, etc) and heating mechanisms (spontaneous emission, straggling, scattering, incoherent space charge, intra-beam scattering, noise instrumentation, etc.) that lead to dissipation, i.e., a non-Liouvillean process. The balance between the two mechanisms yields the equilibrium emittance(s). The existing theories of beam cooling rely on kinetic models and statistical considerations and are, for the most part, successful at describing experimental situations.

From the point of view of thermodynamics and statistical mechanics, *stochastic cooling* is perhaps the most interesting method, as it relies on a scheme reminiscent of the cooling of a gas by Maxwell's demon. In Maxwell's thought experiment, the demon actuates a gate in a hole at a partition between two vessels containing gas (initially in thermal equilibrium). The demon allows the passage of only fast-moving molecules from one vessel to the other and slow ones in the opposite direction. In this fashion, a temperature difference is established that could be used to extract work, as in a perpetual motion machine, in apparent violation of the second law of thermodynamics. In stochastic cooling, detection of the motion of groups of beam particles is used to correct their orbits in subsequent turns in circular accelerators; the net effect is the (irreversible) emittance reduction of the beam by the filling of internal 'voids' in phase space. In this process, the granularity of the beam plays a role in circumventing Liouville's theorem, which is based on a continuous fluid model.

An excellent introduction to stochastic cooling, invented by van der Meer in the early 1970s and first implemented at the CERN Intersecting Storage Rings (ISR), can be found in [13]. The basic scheme has a detector and a kicker separated by a quarter of the betatron wavelength plus integer multiples of the half-wavelength in a ring lattice. In this way, a particle error at the detector, where the betatron amplitude is near its maximum, yields a proportional signal that is sent to the kicker to correct the orbit's slope. Thus, the orbit of the detected particle is corrected over successive turns, but the correction is imperfect mostly because of limited detector bandwidth, amplifier noise, and incoherent effects caused by the actions of other particles. An *upper limit* to the (time) cooling rate can be estimated using a simple model that neglects incoherent effects and assumes ideal detection. If N is the number of particles in a ring, T the revolution period, and W the detector bandwidth, we can write $N_S = N/(2WT)$ for the number of particles per sample passing through the detector (we have assumed equally spaced samples of duration $1/2W$). Further, the corrected kick is $\Delta x = -(\lambda N_S)\langle x \rangle_S$, where λ is a constant, and $\langle x \rangle_S$ is the average sample error. Therefore, with unit gain ($g = \lambda N_S = 1$) at the detector, the cooling rate is $1/\tau_n = 1/N_S$ *per turn*, and the net cooling rate is $(1/\tau_n) \times (1/T)$, in s^{-1}, or

$$\frac{1}{\tau} = \frac{2W}{N},\tag{8.8}$$

which overestimates the optimum cooling rate by a factor of two. The cooling time is best for low-intensity beams. As an example [13], for the Low Energy Antiproton Ring (LEAR), $N = 10^9$, $T = 0.5\ \mu\text{s}$, and $W = 250\ \text{MHz}$, leading to $1/\tau = 0.5\ \text{s}^{-1}$, i.e., a cooling time of 2 s, or four million turns.

8.5 Computer resources

The basic numerical calculations for the linacs and rings discussed are summarized in the Mathcad worksheets **SLAC-3m.xmcd**, **LEDA.xmcd**, **SURF II.xmcd**, and **NSLS-VUV.xmcd**. These files can be downloaded from this book's website. Also available on the website are the MAD-8 and WINAGILE input files for SURF II.

Lastly, the Android application TAPAs (see appendix A) provides an instructive tool for synchrotron radiation calculations for bending magnets, storage rings, and insertion devices.

References

[1] Allen C K *et al* 2002 Beam-halo measurements in high-current proton beams *Phys. Rev. Lett.* **89** 214802

[2] Thomas P W 2008 *RF Linear Accelerators* 2nd Edn (Weinheim: Wiley-VCH)

[3] Murphy J 1993 Synchrotron Light Source Data Book, Version 3.0, October 1993, Brookhaven National Lab., BNL 42333 (revised) Informal Report

[4] Winick H (ed) 1994 *Synchrotron Radiation Sources—A Primer Series on Synchrotron Radiation Techniques and Applications* 1 (Singapore: World Scientific) The article 'Lattices' by Max Cornacchia is especially recommended

[5] Corben H C and Stehle P 1977 *Classical Mechanics* 2nd edn (New York: Dover Publications, Inc)

[6] Bryant P J and Johnsen K 1993 *The Principles of Circular Accelerators and Storage Rings* (Cambridge: Cambridge University Press)

[7] Koutchouk J P 1989 Trajectory and Closed Orbit Correction, CERN LEP-TH/89-2, Geneva https://cds.cern.ch/record/195334?ln=en

[8] David F S 2014 Minimum data acquisition of the equilibrium orbit and tune *UMER Technical Note*

[9] Edwards D A and Syphers M J 2004 *An Introduction to the Physics of High Energy Accelerators* (Weinheim: Wiley-VCH)

[10] Ghodke A and Chou W *Int. Committee for Future Accelerators (ICFA) Beam Dynamics Newsletter No. 44* December 2007 (available online) https://icfa-usa.jlab.org/archive/news-letter/icfa_bd_nl_44.pdf

[11] Möhl D 2005 Beam cooling *CAS CERN Accelerator School* 224–39 CERN-2005-004, 6 June

[12] Sessler A M 1996 Methods of beam cooling *UC-414, Lawrence Berkeley Laboratory* (Berkeley, CA: University of California) February

[13] Möhl D Stochastic cooling for beginners *CAS CERN Accelerator School* 97–161 CERN-84-15

IOP Publishing

A Practical Introduction to Beam Physics and Particle Accelerators (Third Edition)

Santiago Bernal

Chapter 9

Small machines and scaled experiments

In this chapter, we describe four small ring accelerators and examples of Paul trap simulators. We do not include small linacs such as the ones used in medical and industrial applications, or cyclotrons, which are considered small relative to typical large, kilometer-scale accelerators. The first machine is the University of Maryland Electron Ring (UMER), which has been in operation for almost 20 years. We discuss the most salient features of UMER and some of its research accomplishments in the physics of high space-charge (SC) intensity beams in a storage ring.

The second example is the Small Isochronous Ring (SIR), an ion ring originally developed at Michigan State University to study SC dynamics in the isochronous regime.

The third accelerator is the Integrable Optics Test Accelerator (IOTA) proton/ electron ring currently in development and operation at Fermilab National Laboratory. The main thrust of IOTA is the testing of novel nonlinear optics techniques to circumvent resonances and other factors that affect beam quality in electron and hadron rings.

Our fourth ring is the Electron Model for Many Applications (EMMA), the world's first non-scaling fixed-field alternating-gradient (FFAG) accelerator, which was in operation from 2010 to 2012 at Daresbury Laboratory in the UK.

Finally, we consider the Paul trap experiments in the US (the Paul Trap Simulator Experiment (PTSX)), Japan (the Simulator of Particle Orbit Dynamics (S-POD)), and the UK (Intense Beam Experiment (IBEX)) which use tabletop plasma devices to simulate beam dynamics in alternating-gradient (AG) lattices.

The computer resources include *Mathcad* worksheets with basic calculations for all five machines discussed. For UMER, we also illustrate the use of the WINAGILE code (see appendix A) to calculate closed orbits and corrections when magnet errors are present. Additional resources will be posted on the book's website as they are developed.

9.1 The University of Maryland Electron Ring (UMER): a storage ring for space-charge research

The original motivation for UMER was to model beam transport in a recirculator for heavy-ion inertial fusion. Beam transport in such a hypothetical machine entails beam physics phenomena in regimes of high SC intensity, in both transverse and longitudinal dynamics. UMER, completed in 2004 [1, 2] except for the extraction section, employs an AG focusing lattice over a 36-sided polygon inscribed in an 11.52 m circumference. A distinctive feature of the lattice is the high density of its magnets, which allows beam containment from the emittance-dominated to the high-SC-intensity regimes without changing the external focusing (see below, and the discussion in section 4.4). In this section, we describe UMER's lattice and injection and discuss the closed orbit, betatron tunes and resonances, soliton trains and instabilities, and work on nonlinear optics.

9.1.1 UMER lattice and injection

As illustrated in figure 9.1, the main lattice consists of 18 sections, each containing two focusing–defocusing (FODO) cells (four quadrupoles) and two bending dipoles. These air-core elements are based on printed circuits (PCs) which provide the low gradient or field necessary to focus and bend the low-energy electron beam. The chamber at the center of each ring section ('RC' in figure 9.1) houses a fast capacitive beam-position monitor (BPM) and fluorescent screen diagnostics. Three of the ring sections contain glass adapters for induction modules or wall-current (WC) monitors. Single-turn injection of the 10 keV, 20–100 ns electron beam, with a 60 Hz repetition rate, is set up in a direction such that the Earth's magnetic field provides approximately one third of the bending. While the Earth's field helps to bend the beam smoothly around the ring, it also complicates beam alignment, especially on injection. Helmholtz coils that compensate for the Earth's field are used over the straight part of the matching/injection section as well as along the ring chambers; the latter minimize the vertical deflection caused by the horizontal component of the Earth's B-field.

The pulsed injector, also shown in some detail in figure 9.1, consists of two large air-core magnetic PC quadrupoles, a wire-wound dipole, and a number of steering elements (only one is shown in figure 9.1). The quadrupole labeled YQ is rotated (yaw angle) 20° relative to Q5–6; in this way, YQ acts as both Q7 and QR72. The injection dipole PD, on the other hand, is centered at the intersection of three lines: matching line (Q6-PD), RC1 ring line (PD-QR1), and RC18 line (QR71-PD). Injection and steering are implemented such that quadrupole YQ can act as a combined-function element, i.e., for focusing and bending into the ring lattice. The injection dipole, PD, is set up with a bending field that switches polarity before the beam completes one turn. The polarity swing must be asymmetrical because the bending field opposes the action of the earth's B-field on injection but adds to it for circulation.

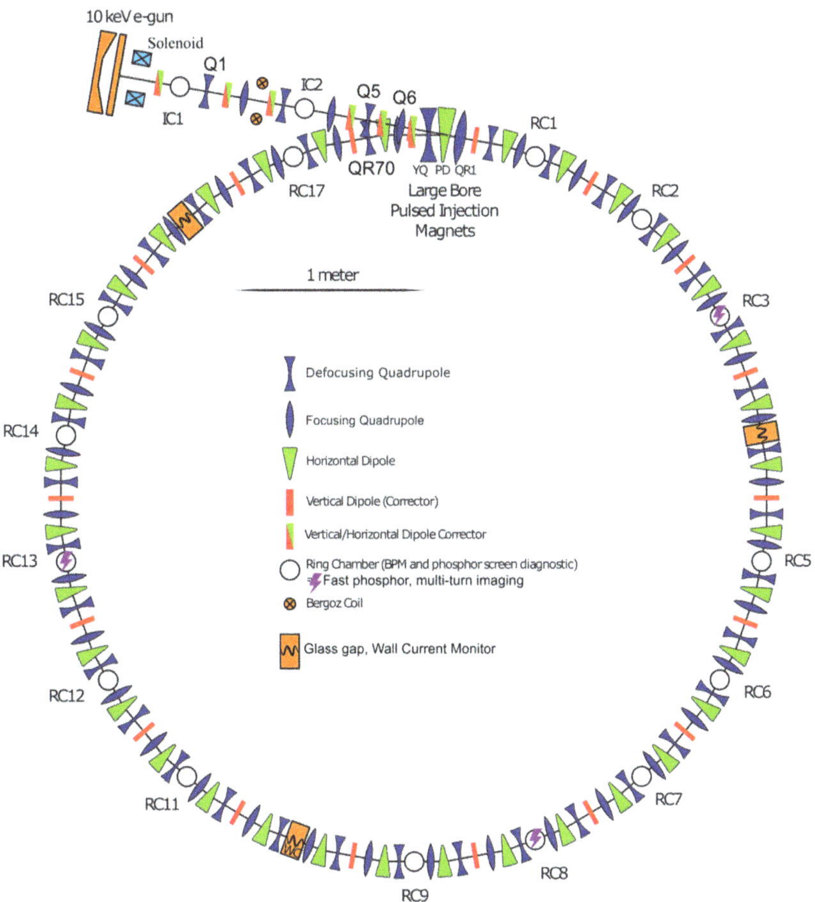

Figure 9.1. Layout of the University of Maryland Electron Ring (UMER). A typical beam bunch is five meters long. Adapted from L Dovlatyan's PhD thesis, by permission [3].

As mentioned above, the high density of quadrupoles in the UMER lattice allows us to transport all beam currents without changing the bare tune. This can be understood from the following approximate equation for the average beam radius in a uniform-focusing approximation of the lattice (see equations (4.31) and (4.32)):

$$a = \frac{S}{\sigma_0}\left(\tilde{\varepsilon}_x \frac{\sigma_0}{S} + K\right)^{1/2} \tag{9.1}$$

where $S = 0.32$ m is the full-lattice period, σ_0 is the 'zero-current' phase advance per period, $\tilde{\varepsilon}_x$ is the 4× root-mean-square (rms), unnormalized emittance, and K is the beam perveance (proportional to beam current). The beam radius 'a' is proportional to $S^{1/2}$ in the regime of emittance-dominated transport, and to S in the limit of strong SC. Thus, it is more efficient to have closely spaced focusing elements for the transport of intense beams, the other parameters being equal, than for the transport of low current.

9.1.2 Closed orbit

One of the first problems in UMER is to define a reference or design trajectory. Following the standard definition, the reference trajectory is determined by the action of the bending dipoles and the earth's B-field on a single particle with zero injection errors (including energy). The net horizontal deflection due to the action of the Earth's B-field (vertical component) is 79.2°, while the combined bending produced by the main dipoles is 280.8°. However, the Earth's B-field acts everywhere, while the dipoles have an effective length of only 3.76 cm. Thus, if the ambient B-field were perfectly uniform around the ring, and there were no mechanical errors, each dipole would have to be powered by the same current (2.35 A), which is approximately 22% lower than the required current (3.0 A) if the Earth's B-field were shielded. In reality, because of the non-uniformity of the ambient B-field and unavoidable mechanical imperfections, the dipoles must be set individually to achieve suitable centroid orbits.

Because of the ambient B-field, the trajectories between bending dipoles (see figure 9.1) are not straight, causing a radial offset of the order of 1 mm relative to the BPM axis at the location of the ring chambers. The beam centroid zigzags between bending dipoles and crosses the quadrupoles off-axis; thus, even in the ideal situation of perfectly aligned quadrupoles, the beam is slightly deflected towards the pipe axis by the focusing quadrupoles and in the outward radial direction by the defocusing ones. Under these conditions, the optimal reference orbit is one with a small offset through the quadrupoles. The closed orbit in a simplified UMER model is studied in an example described at the end of this chapter in the section on computer resources (9.6).

9.1.3 Betatron tunes and space charge

Table 9.1 summarizes the main parameters of UMER. The betatron tunes are determined experimentally (see [4]). For comparison, a simple-minded matrix

Table 9.1. Main parameters of the UMER.

Energy, E	10 keV
Superperiod, N	1
Circumference	11.52 m
Beam current, I_{B}	0.06 to 100 mA
Horizontal betatron tune, ν_{x0}	6.68
Vertical betatron tune, ν_{y0}	6.85
Momentum compaction, α	0.0214*
Tune depression, ν/ν_{x0}	0.15 to 0.92
Coherent tune shift, $\Delta\nu_{\mathrm{coh}}$	0.00036 to 0.61
Emittance (rms, norm.), $\tilde{\varepsilon}_n$	<3.2 μm
Natural hor. chromaticity, ξ_X	−7.31*
Natural vert. chromaticity, ξ_Y	−7.39*

*ELEGANT code, with no SC.

calculation without bending dipoles yields equal bare tunes $\nu_{x0} = \nu_{y0} = 6.83$ (equations (3.16) and (3.17)). If the bending dipoles are included with edge effects, we obtain (ELEGANT code—see computer files in appendix A) $\nu_{x0} = 6.82$, $\nu_{y0} = 6.88$. These edge effects include not only the standard vertical edge focusing performed by the rectangular bending magnets (see chapter 3) but also the field variation at the fringes. An additional effect caused by the Earth's magnetic field in the low-energy UMER can be modeled in ELEGANT through the 'FSE' ('field strength error') parameter. In effect, ELEGANT keeps the reference orbit for 36, $10°$ bending dipoles, while using actual approximately $7°$ dipoles.

The most interesting calculations in UMER relate to the effects of incoherent and coherent SC. We limit our brief discussion here to transverse dynamics effects, although longitudinal SC phenomena also play a significant role in UMER. At the nominal beam current of 0.6 mA, beam transport in UMER is *emittance-dominated*, although the fractional incoherent SC tune shift (equation (4.40)) is sizable, about 14%. The corresponding SC tune depression is $\nu/\nu_{x0} = 0.86$. At the other end of the beam current range, 100 mA, we obtain a fractional incoherent SC tune shift of 85% (!) or an SC tune depression of $\nu/\nu_{x0} = 0.15$. However, UMER operates most of the time with a 6.0 mA beam current; the tune depression in this case is $\nu/\nu_{x0} = 0.64$, which corresponds to strong SC-dominated transport.

The *coherent tune shift* caused by image forces is also appreciable in UMER, despite its low energy. We use equation (4.55) for the case of non-penetrating fields and obtain $\Delta\nu_{\text{coh}} = 0.0036$ for 0.6 mA and 0.61 for 100 mA; at the typical 6.0 mA operating current, $\Delta\nu_{\text{coh}} = 0.035$. The contributions from the term proportional to β^2 in equation (4.55) are small, of the order of 1% for 0.6 mA and negligible for higher beam currents.

9.1.4 Betatron resonances and space charge

In the second edition of this book, we reported calculations of the incoherent tune shift based on the displacement of a half-integer horizontal resonance band of a 10 keV, 6.0 mA electron beam, as observed in a chart of beam losses vs. estimated tunes (figure 7.4 in [5]). For the analysis, we used the coherence resonance theory discussed in section 7.4 of the current edition. However, additional studies by a K Ruisard and L Dovlatyan [3, 6] did not reveal exactly the same resonance band structures as those seen before.

L Dovlatyan studied the resonances for three beam currents: 0.6 mA, 6.0 mA and 20 mA, all at 10 keV. We show below in figures 9.2(a) and (b) charts of beam losses vs. estimated tunes (at the tenth turn) for 0.6 mA and 6.0 mA. The beam loss is the fraction of beam that survives relative to the injected beam, as measured by a WC monitor. Unlike the results reported in [5, 6], no horizontal half-integer resonances are seen in the new experiments at the same tenth turn; instead, a new vertical half-integer resonance is observed at $\nu_{y0} = 7.5$. These differences can be ascribed to improvements in beam steering and matching (which can delay the development of the half-integer resonances) and refined tune measurements. In fact, additional studies by Dovlatyan using numerical analysis of fundamental frequencies (NAFF)

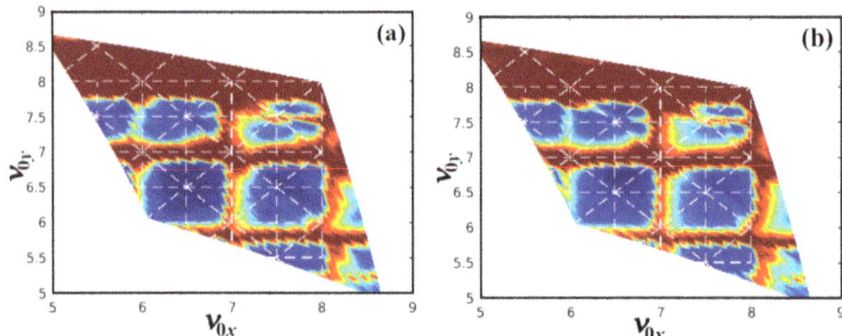

Figure 9.2. Beam losses at the tenth turn as functions of estimated bare betatron tunes for (a) 0.6 mA, and (b) 6.0 mA. Dark red indicates 100% beam loss and blue indicates 100% beam transmission. Adapted from L Dovlatyan's PhD thesis, by permission [3].

[7] techniques to measure tunes revealed the appearance of both horizontal and vertical half-integer resonances for the 0.6 mA beam (at the twenty-fifth turn) and a half-integer horizontal resonance for the 6.0 mA beam (at the fifteenth turn). The different rates of longitudinal expansion (de-bunching) due to SC (see the example in section 6.5) as well as the emittance evolution for the different beam currents complicate the comparison of resonance band structures. The conclusion of these studies is that the integer resonances are not affected by direct SC, as predicted by theory [8], but the expected displacement of the half-integer bands is difficult to ascertain. Additional considerations are given when we discuss the experiments in Paul traps in section 9.5.

9.1.5 Soliton trains and multistream instability

The first observation of soliton wave trains in electron beams was made using UMER in 2012 [9]. Solitons can form in water, EM radiation, plasmas, and other media when the steepening of large wave perturbations is balanced by wave dispersion. When a high-density perturbation is applied at the tail of a long bunch by means of photoemission at the cathode, a fast wave develops into a soliton wave train after two to three turns. The pulses maintain their shapes and the other properties of solitons. The observations have been successfully simulated using the WARP code (see appendix A). Figure 9.3 shows the formation and evolution of a soliton wave train in a 22 mA, 10 keV beam with a 25% perturbation. The initial beam (bottom trace) is measured by a Bergoz current transformer before injection into the ring, while the rest of the traces are taken from the WC monitor in RC10 (figure 9.1).

Without longitudinal focusing, the beam in UMER completely debunches, i.e., it turns into a DC beam that fills the entire ring (see the end of section 6.5). The WC monitor is blind to the DC beam, but partial re-bunching and the onset of an instability are observed after a number of turns. For the 6.0 mA, 10 keV beam, partial re-bunching happens after some 40 turns, and an instability develops after another 40 turns. Filamentation of the beam in longitudinal phase space occurs as the charge wraps around the ring. When the separation in velocity space between filaments is comparable to the sound speed in the beam, the filaments interact and the instability

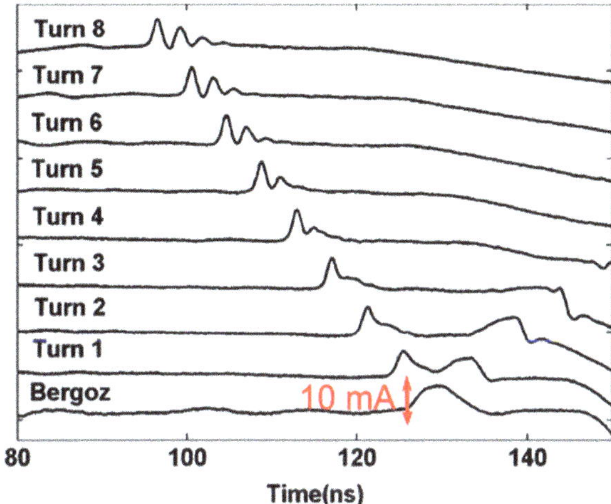

Figure 9.3. Injected and turn-by-turn beam current profiles of a 22 mA electron beam with 25% perturbation. The fast wave on the trailing end of a 100 ns bunch moves to the left and evolves into a soliton wave train. Reproduced with permission from [9] with minor changes.

develops. There is good agreement between theory, WARP simulation, and experiment for different line-charge densities at the center of the bunch [10].

9.1.6 Nonlinear optics

The onset of nonlinear betatron resonances, i.e., those of the third or higher orders, depends on the particle's amplitude (see chapter 7). As a consequence, the fields produced by nonlinear magnets (e.g., sextupoles or octupoles—see appendix B) or errors of nominally linear magnets can lead to a significant reduction of the *dynamic aperture* (DA), defined as the beam size at the limit of stable orbits, at working tunes close to or at the resonance condition (equation (7.7)). In addition, the driving nonlinear fields induce betatron tunes that also depend on amplitude, leading to tune spreads. However, under certain conditions (to be discussed in some detail in section 9.3) the tune spreads induced by nonlinear magnets can mitigate nonlinear resonances without a reduction in the DA. The flexibility of the UMER lattice has facilitated the investigation of these effects in detailed particle-in-cell (PIC) simulations and preliminary experiments with low SC [6, 11].

A modified UMER lattice with ideal three-fold symmetry (a 3.84 m superperiod) was designed to accommodate a 25 cm nonlinear magnet insert between two dipoles in a 20° section; the insert consists of seven short printed-circuit octupoles. The overlap of the octupole fields is tailored to closely follow the profile required to satisfy *quasi-integrable* conditions in combination with the linear optics of the rest of the ring. These conditions allow for 'chaotic but bounded' orbits with moderate tune spreads, unlike the fully integrable systems in which non-chaotic orbits can be maintained with large tune spreads (section 9.3). The beams for the PIC simulations and initial experiments have low SC intensity parameters, corresponding to incoherent tune shifts of 0.005–0.3 at horizontal and vertical bare tunes of 6.8 in the standard UMER lattice. The operating tunes in the modified lattice with the octupole insert, however, are close to 3.13, which leads, according to WARP simulations, to a tune spread of 0.13. The conclusion of these studies is that very tight control of orbital distortion in the octupole section (< 0.2 mm) is required to preserve a sizable DA, a condition not yet achieved in experiments.

9.2 Small Isochronous Ring (SIR): space-charge effects in the isochronous regime

SIR is a low-energy, 6.6 m circumference ion storage ring developed in the early 2000s at Michigan State University in Lansing, MI. Table 9.2 summarizes SIR's main parameters, and figure 9.4 shows the top view schematics of the machine.

Table 9.2. Nominal parameters of SIR [12].

Energy, E	20 keV
H_2^+ beam peak current, I_0	5–25 μA
Circumference	6.57 m
No. of turns	< 200
Bunch length	0.15–5.5 m
Revolution period, T	5 μs
Hor.,vert. bare betatron tunes, ν_{x0}, ν_{y0}	1.14, 1.11
Incoherent SC tune shift at 20 μA, $\delta\nu_{x,y}$	0.005
Momentum compaction, α	1.0

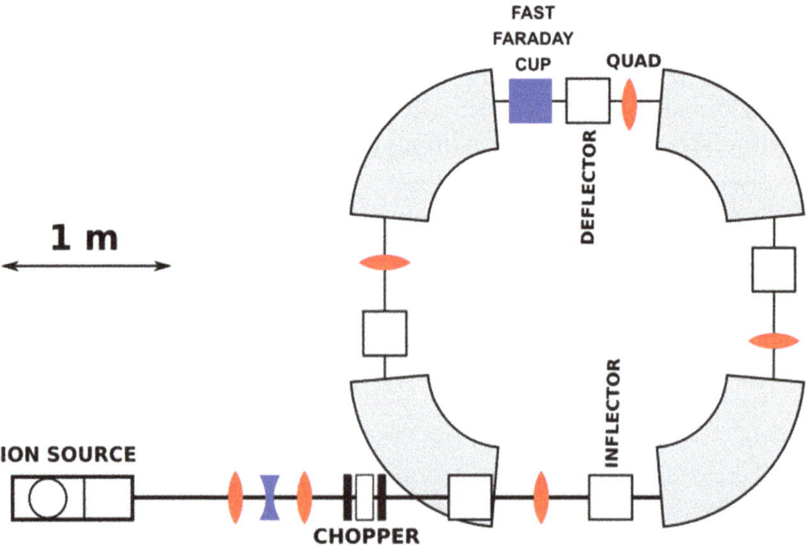

Figure 9.4. Top view schematics of SIR. Adapted with permission from [13].

As shown in figure 9.4, the ring consists of four 90° bend magnets, several dipole magnets and electrostatic quadrupoles, a Faraday cup, and fluorescent screen diagnostics. The Faraday cup provides longitudinal beam profiling with a resolution of 1 ns, while the phosphor screen is used for measurements of the beam cross section. An H_2^+ beam is injected by means of a fast electrostatic inflector that achieves isochronism by proper adjustment of gradient correctors in the bend magnets. As a result of the isochronism, there is no de-bunching from longitudinal SC, but the energy spread increases in the 'frozen' longitudinal distribution. The number of turns is limited due to charge neutralization by electrons in the residual gas. Furthermore, and unlike UMER, the ambient field only plays a minor role, due to the fact that the magnetic rigidity of the SIR beam is some 60 times that of the UMER beam (with both machines at the same energy, e.g., 20 keV).

Ideally, in the isochronous regime, the slip factor η_{tr} defined in equation (3.44) is zero, indicating that the revolution period is independent of energy. Furthermore, at low energy, $\gamma_0 \approx 1$, leading to a momentum compaction factor $\alpha_c \approx 1$. The actual slip factor in SIR is 2×10^{-4} [12].

Research with SIR has revealed a fast-growing SC driven instability, as illustrated by the measured longitudinal bunch profiles shown in figure 9.5. The instability develops over a timescale that is slow compared to betatron oscillations. It can be explained by transverse–longitudinal coupling, whereby deformation of the beam shape ('snaking') leads to a coherent SC field that affects the coherent longitudinal motion. In effect, the deformation of the beam shape enhances a negative-mass-like instability [14–16]. The latter occurs if the slip factor is positive, i.e., greater than the transition energy (section 3.6). However, SC changes the dispersion function

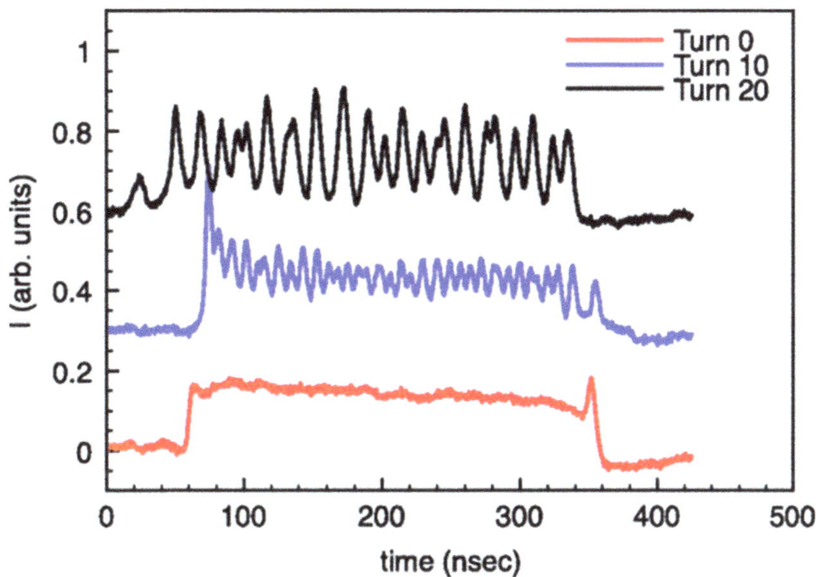

Figure 9.5. Longitudinal bunch profiles in SIR. The beam energy and peak current are 20.9 MeV and 9.3 μA. Reproduced from [12].

(see sections 3.5 and 7.5), which in turn changes the slip factor. From equation (3.43) for the momentum compaction we can write

$$\alpha_{cSC} = \frac{1}{C_0} \int_0^{C_0} \frac{(D_0 + \delta D)}{\rho(s)} ds = \alpha_c + \frac{\delta D}{R_0}, \tag{9.2}$$

where we have included the change in dispersion function that arises from incoherent SC, and R_0 is the average ring radius, $R_0 \equiv \rho_m$. Therefore, from equations (3.44) and (9.2) we obtain (see also [12])

$$\delta \eta = \delta \alpha_{cSC} = \frac{\delta D}{R_0}. \tag{9.3}$$

Pozdeyev *et al* [12] modeled the 'snaking' of the beam by assuming a small distortion and no image forces. Under these assumptions, separate equations for the betatron and centroid motions in a uniform-focusing approximation of the ring lead to a steady-state solution for the centroid part. The net result from the latter is that the change in average dispersion (from its value R_0/ν_0^2—equation (3.38) with $\rho_m = R_0$) is $\delta D > 0$, proportional to $\delta \nu / \nu_0$, the fractional incoherent SC tune shift (the left-hand side of equation (4.38) divided by the bare betatron tune). Therefore, the instability can happen even if the ring is operating below transition.

We can easily show from our discussion in section 7.4 of the effects of linear SC on dispersion that $\delta D / D_0 \cong 2\delta \nu / \nu_0$ for weak SC. We use equations (7.31) and (7.32) with $a = b$, and equation (4.38). Regarding $\delta \nu / \nu_0$, the expression quoted in equation (9) of [12] can be rewritten in SI units as

$$\frac{\delta \nu}{\nu_0} \cong \frac{1}{4\pi\varepsilon_0} \frac{q\lambda}{\gamma^3 m \omega^2 a^2} \frac{1}{\nu_0^2}, \tag{9.4}$$

where λ is the line-charge density, a is the average beam radius, and $\omega^2 = \beta^2 c^2 / R_0^2$; we have used results from appendix A of [13]. Equation (9.4) can easily be shown to be equivalent to our equations (4.38) and (4.39), since $a^2 = \tilde{\varepsilon}_x \bar{\beta} = \tilde{\varepsilon}_x / k_0$, $\nu_0 = R_0 k_0$, and $I = \lambda \beta c$ ($\bar{\beta}$ is the average beta function, not to be confused with the relativistic parameter β).

The growth rate of the instability, on the other hand, can be derived from an expression for the *microwave instability* [15] in the short-wavelength limit of the longitudinal *SC impedance* (see [16] for a definition). At the transition energy, i.e., in isochronous operation, the rate is simply proportional to the peak beam current.

Simulations were conducted that included a PIC solver to calculate the SC fields, with initial phase-space beam distributions based on experimental results. Two regimes were studied, isochronous and below transition. It was found that for a 10 μA peak current, a harmonic of the linear charge oscillations with a wavelength of the order of the beam diameter (1.5 cm) displays the fastest instability growth rate. Furthermore, a simple theoretical model for the instability in the isochronous regime yields growth rates that are some 20% higher than the results obtained from simulations; both theory and simulations, however, predict rates that are much higher than the results obtained using a model of the microwave instability.

Finally, the simulations agree very well with experiments. The latter were conducted in the near-isochronous regime (slip factor $\eta_0 = 2 \times 10^{-4}$) and counted the number of peaks of the longitudinal charge density as a function of the turn number. From a theoretical perspective, the growth rate of the instability is linearly proportional to the beam current, and no dependence on bunch length was observed in the experiments.

9.3 Integrable optics and other physics in IOTA

The IOTA project is part of Fermilab's Accelerator Science and Technology Facility (FAST). The mission of IOTA is to investigate novel nonlinear optics lattices, optical stochastic cooling, SC compensation, and other techniques, all of which are relevant to the intensity frontier of accelerators for nuclear and particle physics.

9.3.1 IOTA lattice and diagnostics

The IOTA ring and its components, shown schematically in figure 9.6, are covered in detail in several recent publications in the *Journal of Instrumentation*. Antipov *et al* [17] conducted a detailed review of the physics program in IOTA and the ideas behind quasi-integrable nonlinear optics [18]. Stancari *et al* [19] discussed the electron lens, a major component of IOTA. Valishev *et al* [20] covered its first experimental results, while Romanov *et al* [21] reported on experiments with a single electron. Most of the material that follows in this section is a summary of the latter references.

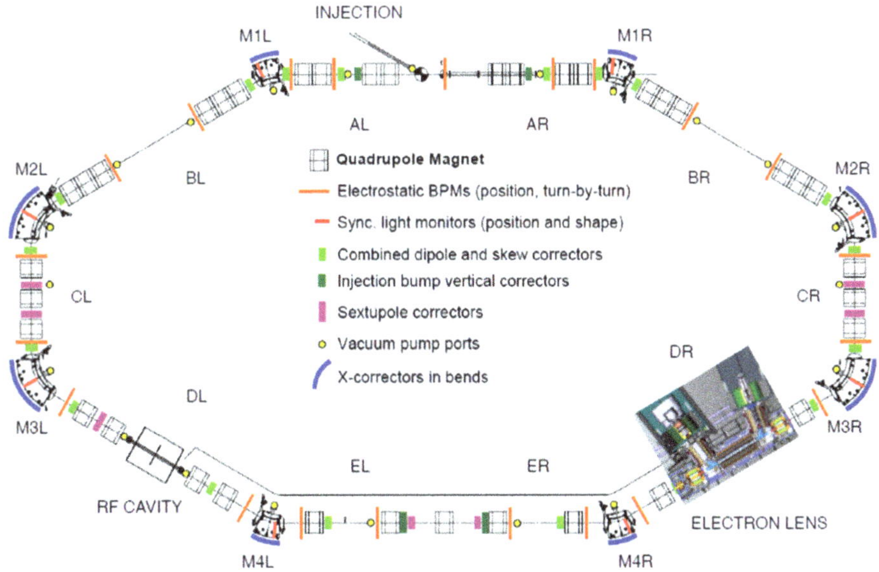

Figure 9.6. Schematic layout of the IOTA ring. The straight sections are labeled AR, BR, etc. (A right, B right, etc.) and AL, BL, etc. (A left, B left, etc). Reproduced from [19] with minor changes.

As shown in figure 9.6, there are eight main bending dipole magnets, 39 quadrupole magnets, 10 sextupoles, and 20 dipole and skew-quadrupole correctors. The 40 m circumference, 50 mm aperture ring is designed for operation with either electron or proton beams. A 150 MeV electron beam is used for integrable optics tests, which require demanding control of the focusing lattice and orbits, with at least a 1% accuracy of the β-functions. This control is feasible with moderately relativistic electrons ($\gamma = 294$), for which SC effects are negligible. For 2.5 MeV protons, on the other hand, SC effects are significant at the IOTA operating parameters. Table 9.3 summarizes the main parameters for both electron and proton beam operations in IOTA.

For electron operations, the IOTA ring is preceded by a 125 m linac consisting of a 5 MeV RF photoinjector, a 50 MeV low-energy beamline, and approximately 100 m of a 300 MeV section. The electron gun is equipped with two solenoids for *emittance compensation* [22] or for the production of angular-momentum-dominated beams (see chapter 5). The low-energy section consists of two superconducting RF cavities, focusing and steering elements, an optional chicane, and a spectrometer dipole that steers the beam into a beam absorber, or lets the beam into the high-energy section when unpowered. The high-energy beamline has eight superconducting cavities that can provide acceleration to 250 MeV. It also contains a FODO lattice and a dogleg to direct the beam into test sections and absorbers, or into the IOTA ring. For proton operation, IOTA's injector is based on a pulsed (1 Hz maximum), 2.5 MeV radio-frequency quadrupole (RFQ). Because of the strong longitudinal SC of the low-energy proton beam, complete de-bunching would occur without RF in less than one turn (see chapter 6). At a pressure of 6×10^{-10} Torr, the lifetime of the proton beam is expected to be 5 min; at 3×10^{-8} Torr, the electron beam lifetime should be 30 min. These lifetimes are calculated at the energies shown in table 9.3.

Diagnostics include capacitive BPMs in the ring (20) and electron injector, with better than 50 μm resolution for a single bunch. In addition, there is a Faraday cup downstream from the RF injector, and wall-current monitors (WCMs) and toroids for absolute calibration of the WCMs. Two WCMs and two toroids are placed

Table 9.3. Nominal parameters of IOTA [17, 19].

Parameter	Electrons	Protons
Max. kinetic energy, E	150 MeV	2.5 MeV
Circumference, C	39.97 m	39.97 m
Revolution period, T	133.3 ns	1.83 μs
Number of particles, peak beam current	2×10^9, 2.4 mA	9×10^{10}, 8 mA
Bunch length	10.8 m	1.7 m
Equilibrium norm. emittance, ε_x, ε_y	0.04, 0.04 μm	0.3, 0.3 μm
RF voltage, frequency, harmonic no.	1 kV, 30 MHz, 4	400 V, 2.18 MHz, 4
Hor.,vert. bare betatron tunes, ν_{x0}, ν_{y0}	4–6	6
Max. incoh. SC tune shift (unbunch., bunched)	$<10^{-3}$	0.5, 1.2

upstream of the high-energy section, and two toroids downstream. (An additional WCM is planned for the ring). There are also ten yttrium aluminum garnet (YAG): Ce crystal screens for transverse beam profiling in the electron injector. The screens are used in conjunction with multislit masks for transverse emittance measurements. Finally, the longitudinal beam diagnostics of the electron beam utilize a streak camera to analyze optical and terahertz radiation from synchrotron and transition radiation.

Special diagnostics for the proton beam are also present, including ionization profile monitors. It should be noted that challenges remain regarding the use of the same RF system and diagnostics (e.g., BPMs) for both electrons and protons. IOTA started operation with electrons in August 2018 and is currently (at the end of 2021) undergoing proton injector installation.

9.3.2 Nonlinear integrable and quasi-integrable optics

The betatron tunes in a circular lattice of (nominally) linear elements, such as quadrupoles, are independent of the betatron amplitude. Beams in such lattices are subject to linear and nonlinear resonances (see chapter 7); the latter arise from nonlinear field components in the actual magnets. In addition, the presence of nonlinear elements such as sextupoles and/or octupoles (see Appendix B) creates amplitude-dependent tunes that may help beam stability but also tend to reduce the DA. It is possible, however, to construct special lattice inserts with nonlinear fields such that the *single-particle* Hamiltonian leads to integrable or quasi-integrable (QI) conditions. In the former case, two integrals of the motion are present: the total energy and a second one, a nonlinear analog of the C–S invariants. For QI conditions, only the total energy is invariant, but the resulting tune spread from the nonlinear fields is expected to enhance beam stability through the mechanism of *Landau damping* [23]. An example of a QI octupole insert in UMER was previously described in this chapter; IOTA has also implemented QI inserts either with a quasi-continuous octupole channel, or with a small number of separated octupole magnets [24]. Furthermore, a fully integrable system based on a 'quasi-continuous nonlinear elliptic-potential focusing channel' is being tested in IOTA. This insert follows the ideas of Danilov and Nagaitsev [25].

As an example of a nonlinear QI potential, we consider an octupole channel. The nonlinear part of the single-particle 2D Hamiltonian, $V(x,y,s)$, can be written as [25]

$$V(x, y, s) = \frac{\kappa}{4\beta(s)^3}(x^4 + y^4 - 6x^2y^2), \tag{9.5}$$

where $\beta(s) \equiv \beta_X(s) = \beta_Y(s)$ is the betatron function, and κ is the octupole strength. The dependence on β ensures that the potential in equation (9.5) is time-independent in normalized phase-space variables, x_N, y_N, ψ, (similarly to the variables defined in equation (7.14)). We write $U(x_N, y_N, \psi) = \beta(\psi)V\left[x_N\sqrt{\beta(\psi)}, y_N\sqrt{\beta(\psi)}, s(\psi)\right]$, so

$$U(x_N, y_N) = \frac{\kappa}{4}\left(x_N^4 + y_N^4 - 6x_N^2y_N^2\right). \tag{9.6}$$

The second invariant in one of the Danilov–Nagaitsev (DN) integrable potentials is a quadratic function of momenta. It involves arbitrary functions of elliptic variables that depend on an arbitrary constant c. The first few terms of the potential for $c = 1$ are [26]

$$U(x_N, y_N) = \frac{1}{2}\left(x_N^2 + y_N^2\right) + t\,\mathrm{Re}\left[\left(x_N + iy_N\right)^2 + \frac{2}{3}\left(x_N + iy_N\right)^4 \right.$$
$$\left. + \frac{8}{15}\left(x_N + iy_N\right)^6 + \cdots \right],$$

(9.7)

where t is the magnitude of the nonlinear part of the potential. Stable small-amplitude motion requires $0 \leqslant t < 0.5$. The corresponding betatron tunes in normalized coordinates are $\nu_{X,\,Y} = \nu_0\sqrt{1\pm 2t}$. For $t \to 0$, the tunes are $\nu_X \to \nu_Y \to \nu_0$, i.e., the linear unperturbed tune, while $\nu_X \to \sqrt{2}\,\nu_0$, $\nu_Y \to 0$ for $t \to 0.5$. Thus, tune spreads of 100% and 40% in the vertical and horizontal planes, respectively, would be possible in principle.

Tests of the nonlinear integrable and quasi-integrable optics (NIO) began in 2019 and continued through 2020 ('run 2'). The QI experiments rely on a 1.8 m insert with 17 discreet octupoles to approximate a continuous potential. A tune spread of 0.04 was achieved, and the observed detuning was in agreement with tracking simulations [20]. The threshold of instability vs. strength of nonlinearity was also measured with the QI insert to test for Landau damping. The instability threshold doubled with the octupole current, clearly showing the stabilizing effect of the tune spread. The runs with the DN configuration, on the other hand, used a specially designed 1.8 m insert with 18 independent magnets of varying apertures. The effect of nonlinearities on betatron resonances as a function of the t parameter was studied; the 100 MeV electron beam could circulate for 110 s at $t = 0.5$, i.e., at integer resonance. An effect of crossing the integer resonance is that the beam splits into two stable beamlets in the vertical plane, in agreement with theory.

Experiments with a nonlinear electron lens are also planned to investigate further integrable optics with both circulating electron and proton beams. The electron lens will be installed in the DR section of IOTA (figure 9.6); the schematics are shown in figure 9.7.

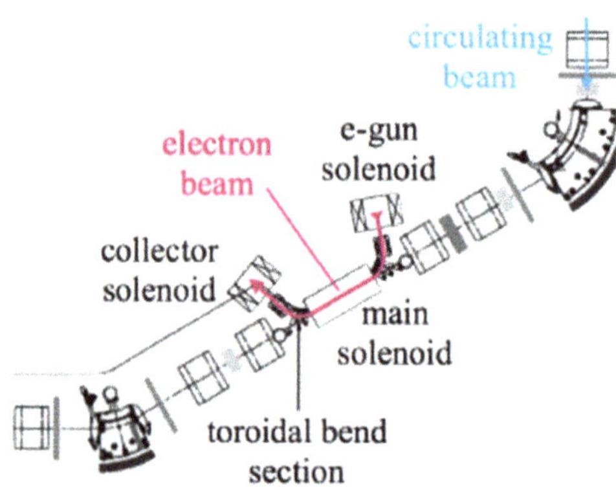

Figure 9.7. General layout of the IOTA electron lens. Reproduced from [17].

The electron lens consists of a low-energy (100 eV–10 keV), high-current electron gun, a beam collector, solenoids at the gun and collector, and a long solenoid (0.7 m) on the main IOTA beamline. Two ideas for NIO are under consideration: (1) a McMillan lens in which the electron beam has a specific current distribution that interacts with the circulating beam to produce 'thin' nonlinear radial kicks that depend on betatron amplitude; (2) an axially symmetric, thick electron lens that will operate in a region of equal and constant betatron functions in the ring. The second option is more robust than the McMillan lens, as it works with any electron-lens current distribution, although the resulting tune spread is smaller than that of the first option. If the thick lens is used, the tune spread can be easily estimated [17, 19] as

$$\Delta\nu = \frac{1}{(B\rho)}\frac{B_z L}{4\pi} = \frac{L}{2\pi\beta_{\text{lat}}}, \tag{9.8}$$

where $(B\rho)$ is the magnetic rigidity, B_z is the long solenoid B-field, L is the solenoid's length, and $\beta_{\text{lat}} \equiv \beta_X = \beta_Y$ is the lattice (constant) beta function in the long solenoid section. For 150 MeV electrons, $(B\rho) = 0.50$ Tm; since $L = 0.7$ m, and assuming $\beta_{\text{lat}} = 4$ m and $B_z = 0.50$ T, we get $\Delta\nu = 0.028$.

9.3.3 Optical stochastic cooling

We saw in chapter 6 how betatron oscillations and the emittance of electron beams are damped by synchrotron radiation (SR) in circular machines. This damping also happens in very-high-energy hadron machines, such as the Large Hadron Collider (LHC). Stochastic cooling (chapter 8), on the other hand, has been employed to improve the luminosity in proton–antiproton machines (e.g., the Intersecting Storage Rings (ISR)) or heavy-ion colliders (e.g., the Relativistic Heavy Ion Collider (RHIC)), but it is impractical in electron rings. Standard stochastic cooling relies on detectors with bandwidths in the radio-to-microwave frequency range, or about 100 MHz to 10 GHz (see equation (8.8) and discussion). In contrast, effective stochastic cooling of electron and positron beams requires bandwidths in the optical frequency range (10^{14} Hz). In optical stochastic cooling (OSC), undulators (chapter 6) are used for both the detector pickup and the kicker. The EM radiation from a particle in the pickup undulator is amplified and gives a correction kick to the same particle in the second undulator. A magnetic chicane is employed to delay the particle to allow time for the amplified radiation to arrive at the kicker at the same time as the particle. IOTA has recently demonstrated 'passive' OSC (2021) of its 100 MeV electron beam with a cooling rate that is one order of magnitude better than natural SR damping. 'Active' OSC, which involves optical amplification, is planned for 2023.

9.3.4 Space-charge compensation and other experiments

Under the right conditions, an electron cloud superimposed on a proton beam can neutralize the effects of SC in the latter (see section 4.6 in [27]). The electron lens discussed above can be adapted to perform SC compensation (SCC) in the IOTA low-energy proton beam. There are two ways to accomplish this. In the first method, the electron gun generates the transverse and longitudinal charge distributions required to reproduce the bunch profiles of the circulating proton beam. In the second method, electrons are generated by ionization of the residual gas to create an 'electron column' that is contained radially by the main solenoid and longitudinally by electrodes, in a configuration similar to a Penning–Malmberg trap [28].

The transverse potential depression for a long charged-particle beam with uniform charge density is given by [27]

$$\phi = \frac{I_b}{4\pi\varepsilon_0\beta c} \cong 30\frac{I_b}{\beta} \text{ (Volts)} , \tag{9.9}$$

where I_b is the beam current and β is the relativistic factor. This yields 3.3 V for the 2.5 MeV, 8 mA IOTA proton beam. Thus, using a single 1 m electron column, the voltage on the electrodes for SCC of the circulating beam in the 40 m ring is estimated to be 130 V. Simulation studies using the WARP code (see appendix A) show that SCC can be optimized by varying the main solenoid field, electrode voltages, and vacuum pressure.

Furthermore, there are plans to do electron cooling of an approximately 2.5 MeV, 0.44 mA proton beam with an estimated cooling time of 20 ms and a reduction of transverse emittance by a factor of ten. It is of special interest to investigate how and if NIO leads to higher brightness when combined with electron cooling. There are also proposals to use the electron beam to generate x-rays, gamma rays, and tunable THz radiation, through several mechanisms; e.g., x-rays from diamond crystals and gamma rays from inverse Compton scattering. We conclude our summary of IOTA/FAST research with a very brief account of an experiment to detect a circulating single electron [21].

We quote from the Fermilab website on IOTA [29]: 'On October 31 [2018], for the first time at Fermilab, we recorded a single electron traveling through an accelerator, the 40 meter-circumference IOTA ring. It circulated for 4 min, corresponding to about 2 billion turns. We set up sensitive diagnostics based on synchrotron radiation, which allowed us to observe individual electrons being lost due to collisions with the residual gas in the vacuum chamber. The very last electron died, poetically enough, on Halloween night at 22:41:25.' Figure 9.8 illustrates the process.

9.4 Fixed-field alternating-gradient accelerators (FFAGs): lessons from EMMA

Particle accelerator science and technology has been dominated for decades by AG synchrotron machines, either for the acceleration of electrons or hadrons or as light sources. However, the limitations of synchrotron rings for new projects in high-energy physics, nuclear physics, and applications in industry and medicine have prompted new ideas, or the revisitation of old ones, for more efficient and economically feasible machines. An example from high-energy physics is the

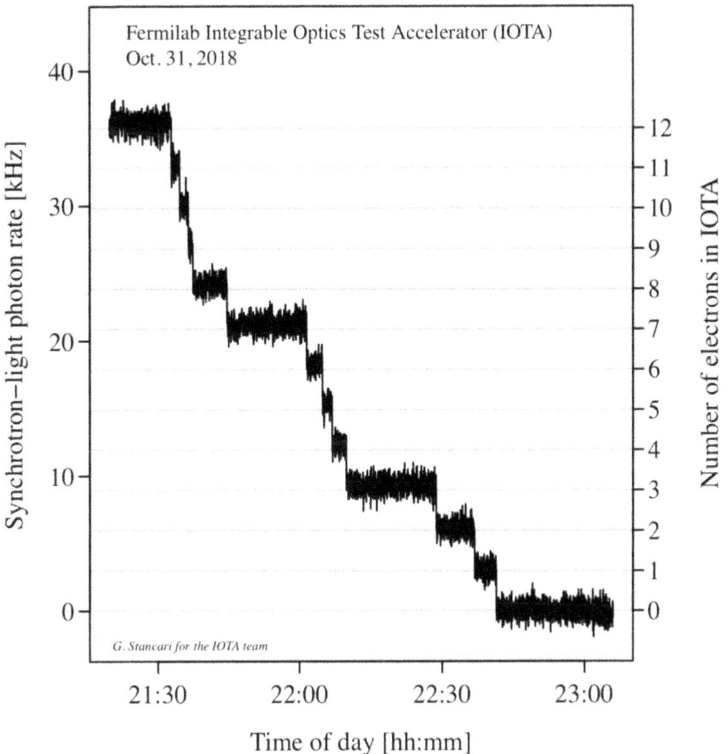

Figure 9.8. Detection of electrons from emitted synchrotron radiation in IOTA. Reproduced with permission from Fermilab [29].

muon–muon collider [30], which requires very rapid acceleration of fast-decaying muons. EMMA was designed as a scaled version of a muon accelerator. EMMA was the first linear non-scaling FFAG to be built and successfully tested. It was operational at Daresbury Laboratory in Cheshire, in the UK, between 2010 and 2012.

Symon and Kerst in the U.S. [31], Ohkawa in Japan, and Kolomensky in the former Soviet Union invented the FFAG concept in the mid-1950s, a few years after the invention of AG (or strong) focusing by Courant, Christofilos, and others. Some of the shortcomings of the original Lawrence cyclotron had been overcome with the synchrocyclotron and the sector-focused cyclotron, and the FFAG would use techniques of the new cyclotrons. See [32] for an introductory treatment and the history of both cyclotrons and FFAGs.

Symon and collaborators from the Midwestern Universities Research Association (MURA) in Wisconsin built several prototype electron FFAGs of the scaling type, but their proposals for proton machines were not approved and FFAGs were abandoned [33]. However, the machines were revisited in Japan in the late 1990s and early 2000s, when several proton accelerators were built as proof-of-principle scaling FFAGs for nuclear physics and medical applications [34].

As the name implies and unlike synchrotrons, FFAGs have magnets that are not ramped during acceleration. This allows faster repetition rates (limited only by RF capabilities), higher average beam current, and larger momentum acceptance. However, the magnets must have large apertures, short lengths, and require special nonlinear B-field radial profiles. In *scaling* FFAGs the magnets are designed so that the particle orbits and tunes are the same at all energies (zero chromaticity), thus avoiding resonance crossing on acceleration. Symon and collaborators [33] introduced two types of scaling FFAG: *radial* and *spiral*. Figure 9.9 illustrates the basic design of a scaling FFAG of the radial type; it consists of eight triplet FDF cells.

The radial B-field (average) profile in scaling FFAGs follows [31, 36]

$$B_{\mathrm{av}} = B_0 \left(\frac{r}{r_0} \right)^k, \tag{9.10}$$

where $B_{\mathrm{av}} = \langle B(\theta) \rangle$ is the azimuthal average over the median plane of the magnet, B_0 is the field at the magnet center, r is the radial distance from the FFAG center to the equilibrium orbit, and k is a constant, $k \gg 0$. The latter can be identified with the negative of the field index n defined in equation (2.26). However, unlike the weak focusing case discussed in section 2.6, the vertical component of the B-field increases with radius. Furthermore, to accomplish both horizontal and vertical orbital stability, a positive-bending radial-sector magnet of field B_{F} is followed by

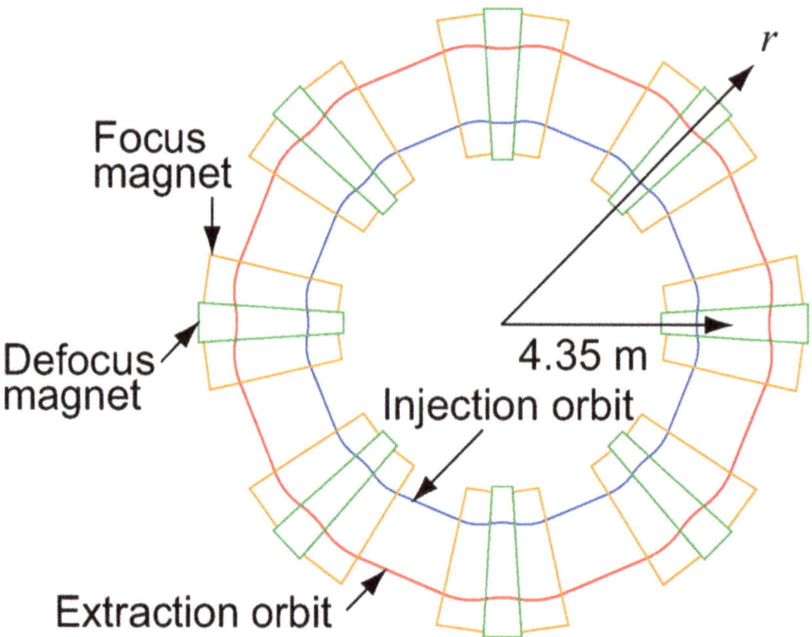

Figure 9.9. Schematics of a radial-sector FFAG. Reproduced, with permission from IEEE [35].

a shorter negative-bending one of field $-B_\mathrm{F}$. To first order, the betatron tunes are given by [36]

$$\nu_r = \sqrt{1 + k}, \quad \nu_z = \sqrt{-k + F^2(1+2\tan^2 \varepsilon)}, \tag{9.11}$$

where $F^2 = \langle (B - B_\mathrm{av})^2 \rangle / B_\mathrm{av}^2$ defines the square of the *magnetic flutter*. For radial scaling FFAGs the angle $\varepsilon = 0$, and the flutter is constant. AG focusing is also realized using edge focusing in the spiral type of scaling FFAG. In these machines, particles enter the magnet faces at an angle ε relative to the normal (see section 2.5), and no alternating polarities are used. To summarize, scaling FFAGs operate with constant tunes from the injection energy to the extracted one, i.e., zero chromaticity, and with self-similar orbits. Additional details can be found in a CERN Accelerator School article by Machida [37].

With fast acceleration (in less than 20 turns) in mind, Mills, Johnstone, and collaborators proposed and designed FFAGs without scaling [38]. *Non-scaling* FFAGs allow the betatron tunes to vary, crossing multiple integer resonances during acceleration. Moreover, high periodicity (N in equation (7.7)) is required to accommodate large momentum acceptance. However, there is more flexibility for lattice design, leading to larger DAs and smaller physical apertures. In the *linear* type of non-scaling FFAG (LNS-FFAG), standard linear quadrupole magnets are used, and the orbital circumference varies quadratically with momentum, enabling the use of a fixed RF frequency at relativistic energies [36].

Near or at the minimum of the orbital circumference as a function of energy, LNS-FFAGs have very small compaction factors, i.e., the machines are nearly isochronous (see section 9.2 on the SIR machine). Furthermore, the Hamiltonian for longitudinal motion has a cubic dependence on energy [36] that leads to stable fixed buckets at two different energies in the energy vs. phase diagram. Thus, it becomes possible to accelerate between buckets instead of inside them, leading to so-called 'serpentine acceleration' as in nearly isochronous cyclotrons

As already mentioned, EMMA is a proof-of-principle LNS-FFAG. It was designed to demonstrate fast and safe crossing of multiple resonances during acceleration and serpentine acceleration. The main parameters of EMMA are summarized in table 9.4, and diagrams of the machine and main cell are shown in figure 9.10.

EMMA's design is described in great (mostly engineering) detail in [41]. The actual research results appeared in a 2012 *Nature Physics* article [39] and also featured on the cover of the journal, which carried a depiction of serpentine acceleration. A good summary of EMMA's results can be found also in a presentation by Machida for the 2014 FFAG workshop at Brookhaven National Lab [42]. The material that follows relies on these three sources.

EMMA's main design feature was its high density of magnets and diagnostics. It had 84 short quadrupole magnets that could be horizontally displaced to produce the dipole component needed to bend the beam around the ring. Thus, eight different lattices with specified F and D quadrupole strengths and displacements were designed. The eight lattices had different cell-tune footprints as a function of

Table 9.4. Nominal parameters of EMMA [36, 39, 40].

Energy, E	10–20 MeV
Circumference	16.57 m
Cells (focusing–defocusing doublets)	42
F quadrupole length	5.88 cm
D quadrupole length	7.57 cm
Rev. time @ 18.5 MeV/c, equiv. momentum	55.3 ns
Radio frequency	1.30 GHz
No. of cavities	19 × 120 kV
Tune shift per cell, ring	$\approx$ 0.3–0.1, $\approx$ 12–4
Normalized transverse acceptance	3 mm
Bunch charge	10–40 pC

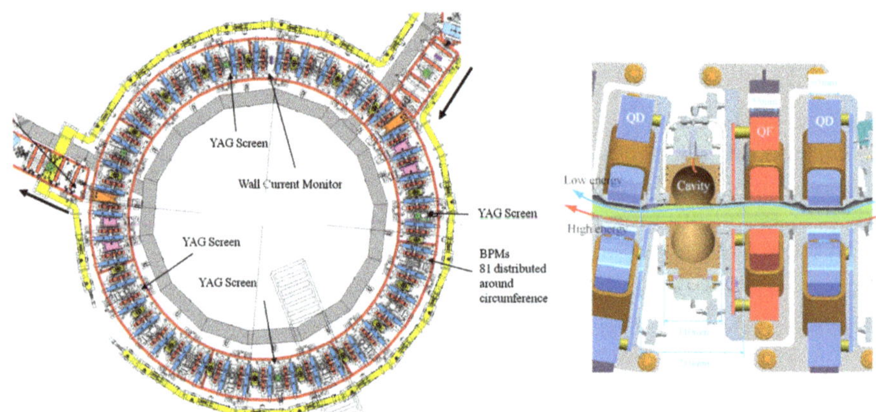

Figure 9.10. Schematics of EMMA and a doublet cell. Reproduced, with permission, from [41].

energy, as illustrated in figure 9.11 for four lattices (b–e) designated 'symmetric' in relation to the parabolic time of flight. Importantly, the quadrupoles in a family should have the same integrated field strength to within 0.01%, according to simulations.

EMMA had a total of 19 normal conducting, single-cell RF cavities, one at every other section except at the points of injection and extraction, with a baseline RF frequency of 1.30 GHz. Other than the obvious use of RF to accelerate the 10 MeV injected electrons, it was also adjusted for time-of-flight measurements at a fixed energy and for resonance crossing studies.

EMMA's diagnostics included 81 capacitive BPMs, two per cell between quadru-poles (except at injection and extraction), and a straight section with a WCM (figure 9.10). The overall resolution of the BPMs was 50 μm. There were also 16 YAG-crystal screens, four in the ring and six in both the injection and extraction beamlines. The latter beamlines had tomography sections for emittance

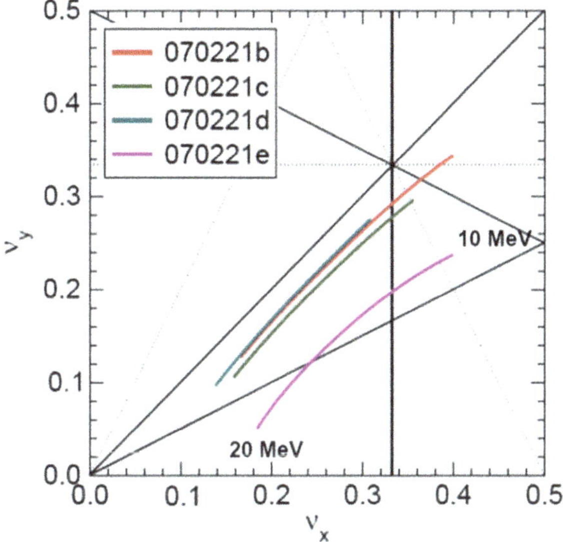

Figure 9.11. Cell-tune footprints for four 'symmetric' lattices. Resonance lines up to the third order are shown. The lattice labeled 'd' does not cross any resonance lines. Reproduced with permission from [41]. (Modified color version from Craddock presentation [36].)

measurements, consisting of three YAG-crystal screens and several quadrupole magnets. The injector comprised the Accelerators and Lasers in Combined Experiments (ALICE) facility: two linacs with superconducting cavities with an acceleration capability of up to 35 MeV. The horizontal acceptance of EMMA was probed by two injection kickers. Particularly challenging were the large angles required for injection and extraction, 65° and 70°, respectively. Additional diagnostics in the original design included Faraday cups at the injection and extraction beamlines, a dipole spectrometer to measure beam energy on extraction, and an electro-optic device for longitudinal profiling of the beam.

Muratori reviewed the commissioning of EMMA in a detailed presentation at the 2011 FFAG Workshop [40]. The machine was achieving more than 1000 turns without RF by August 2010. Acceleration from a momentum of of 12.5 MeV/c to a momentum of 21 MeV/c was achieved in April 2011. At an RF voltage of 100 kV per cavity, i.e., a total of 1.9 MV for all cavities, a wide channel between RF buckets was created for so-called serpentine acceleration for several initial phase values. The orbit and cell tune were obtained as a function of the cell number for five consecutive turns. The results are shown in figures 9.12(a) and (b) for one case of initial RF phase.

The trajectories in longitudinal phase space were determined from beam position and tune measurements. The results, with the momenta obtained from the horizontal beam positions, are displayed for five initial phase values in figure 9.12(c).

In conclusion, EMMA demonstrated serpentine acceleration within six turns from 12 to 18.4 MeV/c, equivalent momentum, a tune variation greater than 4.2 (cell tunes are multiplied by 42—see figure 9.12(b)), and orbital excursions of the order of

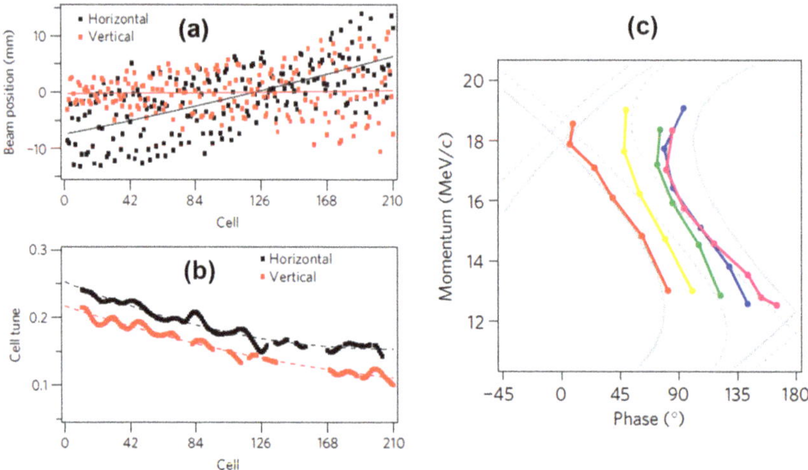

Figure 9.12. (a) Horizontal and vertical beam positions of the accelerated beam over five consecutive turns. (b) Horizontal and vertical cell tunes over five turns. (c) Acceleration in serpentine channels: trajectories in longitudinal phase space for five different values of the initial RF phase. Reproduced, with permission, from [39].

10 mm. The latter figure is well within the 40 mm aperture of the D quadrupoles. Apparently, crossing of integer resonances causes coherent motion without emittance growth, but the momentum spread and lack of chromaticity correction lead to decoherence and emittance growth [42].

9.5 Beam stability and betatron resonances: Paul traps as model accelerators

A Paul trap, invented by Wolfgang Paul and collaborators around 1955, can confine ions in three dimensions by means of a transverse RF quadrupole field and axial DC fields. Okamoto *et al* in Japan (1999) and Davidson *et al* in the US (2000), working independently, realized that the equations governing the dynamics of ions in a *linear Paul trap* (LPT) were isomorphic to those describing charged-particle beams in an AG lattice. In effect, 'the transverse RF quadrupole focusing in the laboratory frame (the time-domain) has the same property [character] as the discrete magnetic AG focusing in the beam frame (the s-domain)' [43]. Thus, an LPT can be used to model AG systems, permitting the simulation of beams in almost any type of lattice geometry and at a very low cost. However, LPTs have some limitations: with fixed electrode lengths, modeling is restricted to 2D (no end effects), and no dispersive effects can be studied because of the lack of 'bending' elements.

Gilson, Davidson, and collaborators presented the results of experiments with the Paul Trap Simulator Experiment (PTSX) at the Princeton Plasma Physics Laboratory in a 2004 *Physical Review Letters* publication [44], and additional details were presented in a later paper [45]. Okamoto and colleagues, on the other hand, worked extensively to simulate beam stability and resonances, with various degrees of SC intensity, using much smaller linear LPTs named S-POD (Simulator of Particle Orbit Dynamics) [46], later collaborating with groups in the UK

(IBEX—Intense Beam Experiment). We illustrate PTSX and S-POD/IBEX in figures 9.13(a) and (b), and summarize their parameters in table 9.5.

If $\omega_0 = \omega_q$ represents the angular frequency of transverse focusing in a quadrupolar LPT, and f is the RF frequency, then the bare phase advance per RF cycle is $\sigma_0 = \omega_0/f$, and the corresponding betatron bare tune is $\nu_0 = \sigma_0/2\pi$. In the LPT [44, 48],

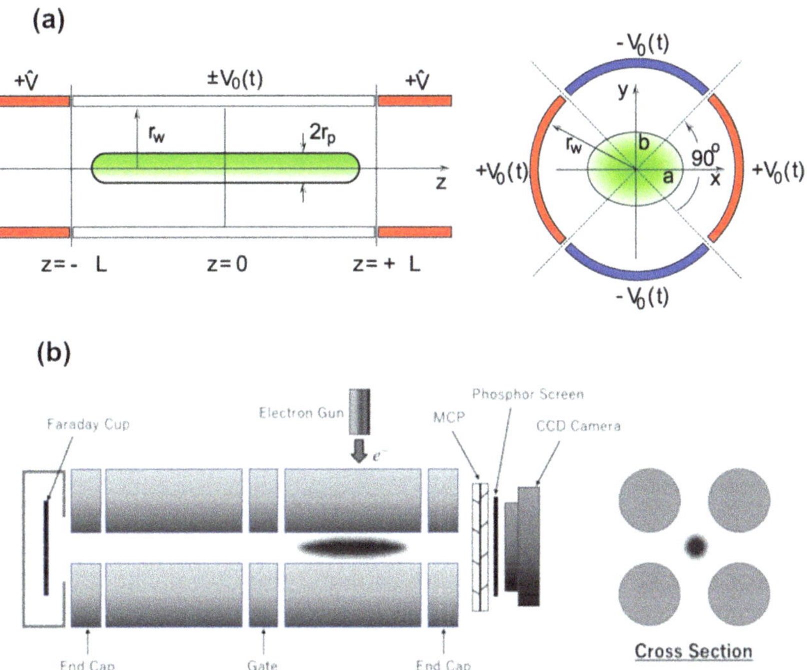

Figure 9.13. (a) PTSX device. Reproduced, with permission, from [44]. (b) S-POD device. Reproduced, with permission, from [46]. The IBEX trap is similar to S-POD. See table 9.5 for parameter values.

Table 9.5. Main parameters of three LPT devices used for accelerator physics research [45, 47].

	PTSX	S-POD II-III	IBEX
Length, $2L$	2.0 m	0.20 m	0.075 m
Ion species, m_b (amu)	C_s^+, 133	$^{40}Ar^+$, $^{40}Ca^+$	$^{40}Ar^+$
RF frequency, f	75 kHz	1 MHz	1 MHz
Voltage amplitude, V_0	235 V	< 100 V	< 100 V
Trap radius, r_w	10 cm	5 mm	5 mm
Typical plasma rms radius, r_p	2.7 cm	1 mm	1 mm
Lowest tune depression, σ/σ_0	0.45	0.8	0.8
Temperature, $k_B T$	≈ 0.5 eV	≈ 0.25 eV	≈ 0.25 eV
Typical trapping time, t_t	> 100 ms	> 100 ms	> 100 ms

$$\omega_0 = \frac{8eV_0}{\pi m_b r_w^2 f}\xi, \quad \xi = \frac{\sqrt{2}}{4\pi}, \tag{9.12}$$

where the expression for ξ is valid for a *sinusoidal* RF waveform, V_0 is the voltage amplitude, and r_w is the trap radius. Therefore, the bare 'cell' tune in the case of a sinusoidal waveform is

$$\nu_0 = \frac{\sqrt{2}}{\pi^3}\frac{eV_0}{m_b}\left(\frac{1}{f\,r_w}\right)^2. \tag{9.13}$$

Gilson *et al* [44] used a 'normalized intensity parameter' $\hat{s} = \omega_p^2(0)/2\omega_q^2$ to quantify the SC intensity in PTSX. The numerator in the expression for $\hat{s}$ is the (on-axis) plasma frequency squared, and the denominator contains, in the uniform-focusing approximation, the average transverse focusing angular frequency squared. It is straightforward to see that $\hat{s}$ exactly corresponds to our SC intensity parameter χ in equation (4.28), since $K = \omega_p^2 a^2/2v^2$, $\omega_q^2 = \omega_0^2 = v^2 k_0^2$, which is also the case in the smooth approximation. Furthermore, the plasma 'global radial force-balance equation' can be translated into beam parameters. The plasma version, equation (4) in [44], is

$$m_b \omega_q^2 R_b^2 = 2k_B T + \frac{N_b e_b^2}{4\pi\varepsilon_0}, \tag{9.14}$$

with $R_b = a$, beam radius, k_B = the Boltzmann constant (not to be confused with the wavenumber k), and N_b is the line particle density (in m^{-1}). The smooth-approximation (beam) envelope equation (4.25), using also equation (4.26), is

$$k_0^2 a - \frac{K}{a} = \frac{\tilde{\varepsilon}_x^2}{a^3}, \tag{9.15}$$

which can be seen to correspond to (plasma) equation (9.14) if we realize that $\omega_0^2 = v^2 k_0^2$, $\tilde{\varepsilon}_x^2 = k_B T a^2/m_b v^2$, and $N_b = 2\pi a^2 n_b$ and use the expression for ω_p implicit in equation (4.12) in terms of the volume density n_b (in m^{-3}).

The line density N_b and plasma radius R_b are calculated from charge profile data obtained using a movable Faraday cup at the end of PTSX. Plotting the left-hand side of equation (9.14) vs. the second term on the right-hand side yields a straight line for small values of N_b; thus, a transverse temperature $k_B T = 0.5$ eV is inferred. Large values of N_b, however, yield a departure from linearity, apparently because of a mismatch between the sizes of the ion source and the plasma. In any case, at an SC intensity of $\chi = 0.18$, no significant radial profile changes are observed in the plasma over trapping times of more than 300 ms. This corresponds to 300×10^{-3} s $\times$ 75 kHz $= 22.5$k cycles, or 22.5 km, if the lattice full-period of the equivalent AG system is 1 m. Figure 9.14 illustrates the evolution of the radial profile.

Gilson *et al* [49] reported an observation of envelope instability (see section 7.4) at $\sigma_0 = 90°$, which was obtained with $V_0 = 425$ V, $f = 75$ kHz (equation (9.13)) with sinusoidal trapping. Additional work with PTSX involved a study of beam

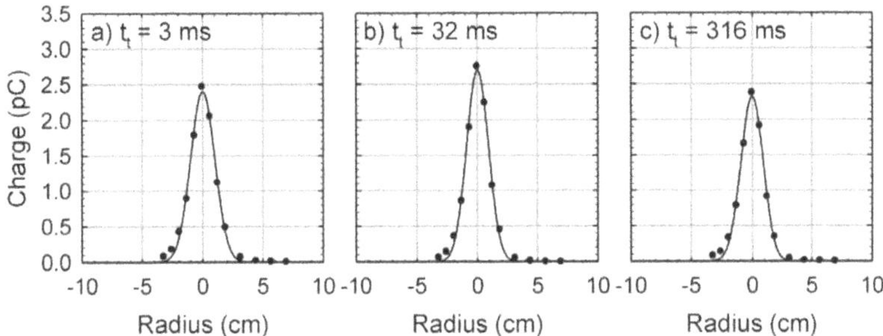

Figure 9.14. Evolution of the radial charge profile in PTSX for an SC intensity of $\chi = 0.18$. Reproduced, with permission, from [44].

degradation, as measured by on-axis charge decay in the plasma induced by random noise applied to the trap voltage [50]. The noise was implemented in such a way as to mimic quadrupole gradient errors in an AG system; it was found that the transverse emittance increases in an approximately linear fashion with the noise duration for noise levels of up to 1.5% of the voltage amplitude. The study was complemented with simulations performed using the WARP code (see appendix A).

Investigations with S-POD at Hiroshima University and with IBEX at the University of Oxford and Rutherford Appleton Laboratory have concentrated on experiments and simulations of betatron resonances with various SC intensities and lattice configurations (i.e., RF waveforms). The Hiroshima University group has had a very good publication record on this complex subject over the last few years. Before summarizing some of their work, we discuss the main points of a recent paper by K Kojima, H Okamoto, and Y Tokashiki [51] and the interesting controversy that it triggered. The latter is represented by a comment from Hofmann [52], a well-known expert from the Gesellschaft für Schwerionenforschung (GSI) in Germany, and the reply by Kojima *et al* [53].

Kojima *et al* proposed to use an expression similar to our equation (7.18) to empirically derive conditions for both 'coherent' (i.e., 'collective') and 'incoherent' (i.e., 'single-particle') resonances. In their notation we have,

$$k(\nu_{0x} - C_m\Delta\bar{\nu}_x) + l(\nu_{0y} - C_m\Delta\bar{\nu}_y) = \frac{n'}{2}. \tag{9.16}$$

In this equation, $\Delta\bar{\nu}_{x,\,y}$ represent *rms betatron tune shifts*, C_m is the 'coherent mode coefficient' introduced in equation (7.19); k, l are integers such that $|k| + |l| = m$, the order of the resonance, and the right-hand side includes the extra factor of 1/2 that we associated in chapter 7 with envelope instability (see also the 1D version, equation (7.22)). Our first observation is that, apparently, Kojima and collaborators do not interpret the extra factor of 1/2 as being simply related to the envelope instability, but a general statement that in effect *doubles* the number of resonance stopbands for *all* orders m. The second observation is that the same semi-empirical coefficient C_m is employed for both transverse planes and it depends only on the order of the resonance, unlike e.g., our equation (7.24). In their words, 'The semi-empirical rule proposed in [equation (9.16)] for the construction of a stability chart includes the incoherent aspect of resonances as a part. The two essentially different aspects, i.e., coherent resonances in the core and incoherent resonances in the tail, should suffice to describe the stability of the whole beam initially matched to the lattice.' [53]. Furthermore, 'the proposed semi-empirical formula in [equation (9.16)] enables one to plot resonance lines of practical importance very quickly and easily without solving a set of complicated differential equations numerically. It predicts the locations of instability bands under arbitrary initial beam conditions. All we need to know in advance are the basic lattice design plus only the rms tune depression that can readily be calculated from the rms envelope equations.' [53]. Figure 9.15 illustrates the concept for a revised tune chart.

Hofmann [52], on the other hand, does not think that the standard tune charts, which are based on the incoherent picture, should be revised. According to Hofmann, there is limited experimental evidence from real accelerators of second-order coherent resonances or instabilities, and no observations of higher than second-order coherent or parametric resonances.

There is also disagreement on the interpretation of the solutions of Vlasov's equation [27] for coherent vs. incoherent resonances. The 2019 paper by Kojima *et al* asserts that Vlasov's theory precludes the existence of incoherent resonances in the

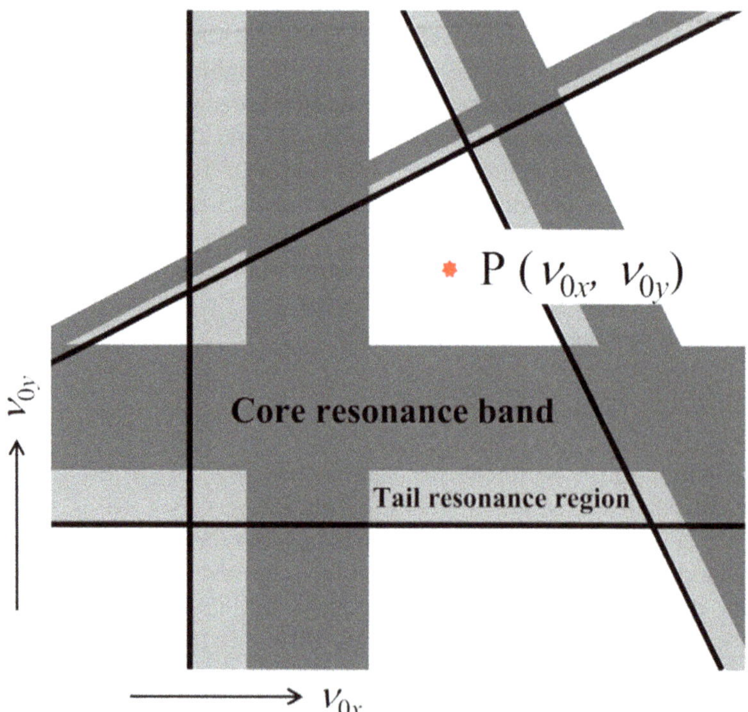

Figure 9.15. Conceptual illustration of the desired operating point in a bare-tune chart relative to the resonance stopbands derived from equation (9.16). Reproduced from [51].

beam core, but Hofmann says that the incoherent spectrum is already taken out of the picture as part of a mathematical simplification towards solving Vlasov's equation. Furthermore, the doubling of the stopbands (the factor of 1/2 in equation (9.16)) is, according to Hofmann, only valid to second order. It would not work for higher orders because of Landau damping in the Gaussian-like beams treated by Kojima *et al.* Finally, experiments with the GSI Schwer-Ionen-Synchrotron (SIS)-18 in Germany and with the proton synchrotron (PS) at CERN on third-order resonances (driven by sextupole errors) do not seem to agree with the empirical framework of Kojima *et al.* The latter defend their work by first emphasizing the simplicity and flexibility of S-POD experiments (and simulations with the code WARP), which have shown the existence of 'linear and low-order nonlinear coherent instabilities [and resonances].' As for comparison with experiments in real accelerators, Kojima *et al* present a case based on the proton rapid cycling synchrotron (RCS) at the Japan Proton Accelerator Research Complex (J-PARC) and reexamine the analysis of the high-intensity beam experiments in SIS-18.

To conclude this section, we mention work by the IBEX group in the UK [47], and additional research by the Hiroshima University group. Martin and collaborators studied four different uncoupled resonances at cell bare tunes equal to $\nu_0 \equiv \nu_{0x} = \nu_{0y} = 1/8, 1/6, 1/5,$ and $1/4$, measured the rms tune shifts, evaluated the A_m factors in the equation

$$\nu_0 - A_m \Delta\bar{\nu} = \frac{1}{2}\frac{n}{m}, \tag{9.17}$$

and compared the A_m values with the theoretical C_m values (from K–V distribution analysis), equation (9.16) with $m = k = l$, and $n = n'$. The WARP code was used to simulate the experiments and evaluate rms emittance growth over a small fraction (550) of the actual number of focusing periods in the experiment (10^5); Gaussian particle distributions were employed. As shown in figure 9.16, the resonances shift to higher bare tunes for increasing numbers of ions in the LPT. The tune shifts vary between 0.015 (at $\nu_0 = 1/4$) and 0.030 (at $\nu_0 = 1/8$) at the maximum ion number of 10^7; the corresponding SC intensity parameters (equation (4.38)) vary between $\chi = 0.120$ and 0.480. Of particular interest is the resonance at $\nu_0 = 1/4$. A strong second-order parametric resonance (envelope instability), $n = 1$, $m = 2$ in equation (9.17), overlaps with the fourth-order incoherent resonance at $n = 2$, $m = 4$; the competition between these two had been studied in simulations by Hofmann and Boine-Frankenheim [54].

The slopes in the plots of $\Delta\bar{\nu}_{\exp}\varepsilon_{\mathrm{rms}}$ vs. the trapped ion number N provide a measure of the coefficients A_m, since we can write $\Delta\bar{\nu}_{\exp} = A_m\Delta\bar{\nu}_{\mathrm{KV}} \propto A_m N/\tilde{\varepsilon}_x$ for the observed resonance shifts (figure 9.16) using equations (9.17) and (4.39) (note the difference between the experimental and theoretical $\Delta\bar{\nu}$). The result is $A_\mathrm{m} = 0.517 \pm 0.091$ for the 1/4 resonance, which, in our opinion, would correspond to the symmetric envelope mode discussed following equation (7.21). However, Martin

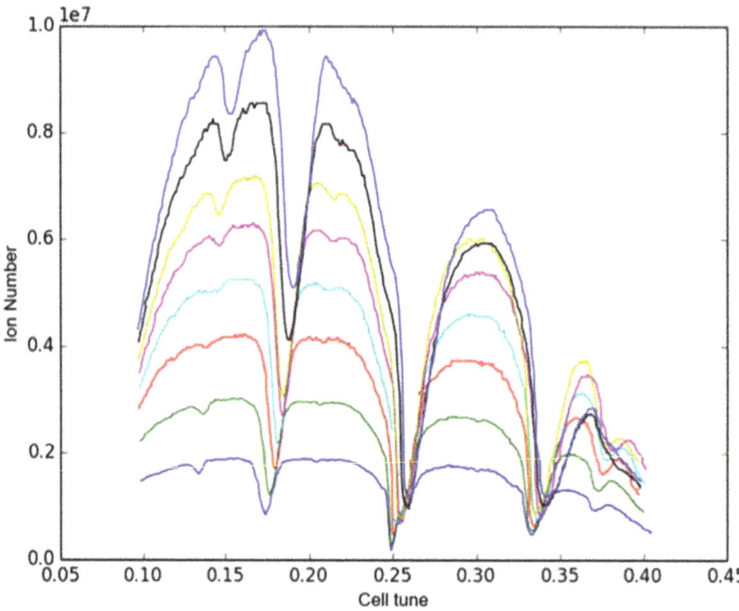

Figure 9.16. Extracted number of argon ions vs. bare cell tunes in experiments by the IBEX group. Each point represents one experiment. Reproduced from [47].

et al compare their result, for reasons that are unclear at the time of writing, to the coefficient for the antisymmetric mode, i.e., 3/4. For the resonances at $\nu_0 = 1/8$, 1/6, on the other hand, the calculated coefficients A_m are larger but close to one, as expected for 'incoherent' resonances. The case $\nu_0 = 1/5$ yields $A_m \approx 0.8$, compared to the theoretical value of one. No further analysis of this case is presented. The source of the nonlinear resonances, i.e., $m > 2$, could be the nonlinear field components caused by the cylindrical electrodes of the LPT; an ideal quadrupolar LPT would use hyperbolic electrodes. Furthermore, some of the discrepancies between theory and experiment can be ascribed to the use of Gaussian or Gaussian-like particle distributions in experiments and WARP simulations vs. calculations of C_m based on the K–V distribution. The resonances at $\nu_0 = 1/6$, 1/4, and other tunes have also been studied by the Hiroshima group [55, 56] and reviewed by Okamoto *et al* [46]. Their conclusion for $\nu_0 = 1/6$ is that it corresponds to an $m = 3$ mode (equation (9.17)) instead of a sixth-order resonance; at $\nu_0 = 1/4$, beam quality degrades significantly, even for fast resonance crossing. A scaling law for emittance growth is derived for the latter case.

Because of their simplicity, flexibility, and compactness, LPTs are elegant devices for modeling beam dynamics in particle accelerators with many different AG lattice geometries. For example, the 42-fold symmetric lattice of the EMMA ring (figure 9.10) has been simulated in an LPT to study integer resonance crossing [56–59]. Furthermore, the 'ultimate control of tune depression' using laser cooling of $^{40}\mathrm{Ca}^+$ ions in an LPT has been proposed to produce the equivalent of crystalline beams [58]. Other proposals include the construction of multipolar ion traps to excite nonlinear resonances in a controllable way [60] and to simulate IOTA's lattices with quasi-integrable nonlinear optics [61, 62]. All these ideas prove that LPTs and (nonlinear Paul Traps) NL-PTs have a bright future in basic research into beam dynamics in accelerators.

9.6 Computer resources

The book's website contains MATHCAD worksheets for all five machines described in this chapter. The calculations aim to reproduce entries in the parameter tables (see the text), such as incoherent and coherent SC tune shifts.

The *WINAGILE* lattice file **ring-ex2.lat** represents an idealized model of the UMER machine with 36 bending dipoles and 72 quadrupoles (see figure 9.1) and a constant external B-field of -0.4 G (pointing downwards). The latter is approximated in the kick file **wina-ex3.kck**, which specifies constant kicks of 1–2 mrad over drifts of about 1–2 cm. In addition, there is a constant kick (6.1 mrad) applied over the effective length of each quadrupole, and positive vertical kicks (18 mrad) applied at each sector bend (SBEND) element (the bending dipoles are represented by two-sector, 5^0 bends and a zero-length *corrector* magnet between them). The effect of the positive kicks is to simulate the reduced field of the dipoles that is required in the presence of the assumed constant -0.4 G Earth's B-field. The magnet *random errors* implemented are: $\sigma = 0.2$ mrad, bending dipole kick error; $\sigma = 2.0$ mrad, transverse tilt error of dipoles; and $\sigma = 0.2$ mm, transverse shift error of quadrupoles. The error distribution is truncated at 2.2σ.

Figure 9.17 displays three horizontal closed orbits (COs) over one turn (11.52 m or 36 FODO periods): the ideal CO (with the constant Earth field), the CO with magnet errors, and the corrected CO. Twenty dipole correctors are employed along with 76 monitors (all quadrupoles are used as virtual BPMs).

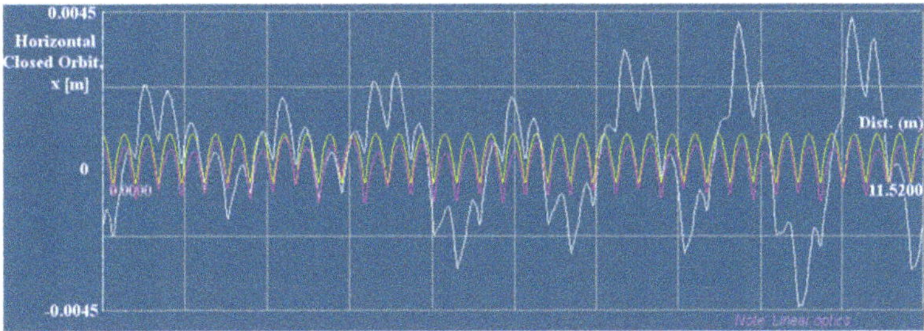

Figure 9.17. Three WINAGILE one-turn horizontal closed orbits in an ideal UMER model (figure 9.1). In all three cases, there is a constant vertical B-field equal to -0.4 G. The yellow trace represents the closed orbit in the absence of magnet strength and alignment errors. The white trace represents the closed orbit in the presence of errors (see text), and the purple trace represents the corrected closed orbit. The WINAGILE files and instructions that can be used to reproduce these results are available online.

References

[1] Bernal S *et al* 2005 Commissioning of the University of Maryland Electron Ring (UMER) *2005 Particle Accelerator Conference (Knoxville, TN)* (Piscataway, NJ: IEEE) FOAD005

[2] Bernal S *et al* 2006 Space-charge beam physics research at the University Of Maryland Electron Ring (UMER), *Proc. of 39th ICFA Advanced Beam Dynamics Workshop, HB2006 (Tsukuba, Japan)* 218–22

[3] Dovlatyan L 2020 *PhD Thesis* University of Maryland, College Park http://hdl.handle.net/1903/26821

[4] Bernal S, Beaudoin B, Cornacchia M and Sutter D 2013 Stability of emittance vs. space-charge dominated beams in an electron recirculator, *Proc. of PAC2013 (Pasadena, CA)* TUPAC31 https://accelconf.web.cern.ch/pac2013/papers/tupac31.pdf

[5] Bernal S 2018 *A Practical Introduction to Beam Physics and Particle Accelerators* 2nd edn (San Rafael, CA: Morgan & Claypool Publishers)

[6] Ruisard K 2018 *PhD Thesis* University of Maryland, College Park http://hdl.handle.net/1903/20942

[7] Laskar J, Froeschlé C and Celletti A 1992 The measure of chaos by the numerical analysis of the fundamental frequencies. Application to the standard mapping *Physica D* **56** 253–69

[8] Baartman R 1998 Betatron resonances with space charge *AIP Conf. Proc.* vol 448 (Woodbury, NY: AIP Publishing) pp 56–72

[9] Mo Y C, Kishek R A, Feldman D, Haber I, Beaudoin B, O'Shea P G and Thangaraj J C T 2013 Experimental observations of soliton wave trains in electron beams, *Phys. Rev. Lett.* **110** 084802

[10] Beaudoin B L, Bernal S, Haber I, Kishek R A and Koeth T 2013 Experimental observations of a multi-stream instability in a long intense beam, *Proc. of the 2013 Int. Particle Accelerator Conf. IPAC (Shanghai, China)* https://s3.cern.ch/inspire-prod-files-c/cd3b55c334324e19b7b4a7bf4f42bff4

[11] Ruisard K, Komkov H B, Beaudoin B, Haber I, Matthew D and Koeth T 2019 Single-invariant nonlinear optics for a small electron recirculator, *Phys. Rev. Accel. Beams* **22** 041601

[12] Pozdeyev E, Rodriguez J A, Marti F and York R C 2009 Longitudinal beam dynamics studies with space charge in small isochronous ring *Phys. Rev. ST. Accel. Beams* **12** 054202

[13] Pozdeyev E 2003 CYCO and SIR: new tools for numerical and experimental studies of space charge effects in the isochronous regime, *Dissertation*, Department of Physics and Astronomy, Michigan State University. Available from https://groups.nscl.msu.edu/nscl_library/Thesis/Pozdeyev,%20Eduard.pdf

[14] Wiedemann H 2007 *Particle Accelerator Physics* 3rd edn (Berlin: Springer) p 692

[15] Zotter B 2013 Single-bunch instabilities in circular accelerators *Handbook of Accelerator Physics and Engineering* ed A W Chao, K H Mess, M Tigner and F Zimmermann 2nd edn (Singapore: World Scientific Publishing Company) 2.4.9 pp 147–9

[16] Zotter B 2013 Space charge effects in circular accelerators *Handbook of Accelerator Physics and Engineering* ed A W Chao, K H Mess, M Tigner and F Zimmermann 2nd edn (Singapore: World Scientific Publishing Company) 2.4.5 pp 137–9

[17] Antipov S *et al* 2017 *JINST* **12** T03002

[18] Antipov S A *et al* 2017 *JINST* **12** P04008

[19] Stancari G *et al* 2021 *JINST* **16** P05002

[20] Valishev A *et al* 2021 First results of the IOTA ring research at Fermilab *12th Int. Particle Acc. Conf., IPAC (Campinas, SP, Brazil)* https://cds.cern.ch/record/2783805/files/document.pdf

[21] Romanov A *et al* 2021 *JINST* **16** P12009

[22] Ferrario M 2013 Space-charge dominated beams in guns and transport lines *Handbook of Accelerator Physics and Engineering* ed A W Chao, K H Mess, M Tigner and F Zimmermann 2nd edn (Singapore: World Scientific Publishing Company) 2.4.4 pp 133–6

[23] Hofmann A 2006 *Landau Damping, in CERN Accelerator School Proceedings* CERN 89-01 http://cds.cern.ch/record/196787/files/CERN-89-01.pdf

[24] Valishev A *et al* 2020 Nonlinear Integrable Optics (NIO) in IOTA Run 2, Fermilab IOTA/ FAST Experiment Report—Beams-doc-8871—November 19 https://beamdocs.fnal.gov/AD/ DocDB/0088/008871/001/IOTA_NIO_Run2_Report.pdf

[25] Danilov V and Nagaitsev S 2010 Nonlinear accelerator lattices with one and two analytic invariants *Phys. Rev. ST Accel. Beams* **13** 084002

[26] Nagaitsev S, Valishev A and Danilov V 2010 Nonlinear optics as a path to high-intensity circular machines, *Proc. of the 46th ICFA Advanced Beam Dynamics Workshop on High-Intensity and High-Brightness Hadron Beams, HB2010 (Morschach, Switzerland)* https:// accelconf.web.cern.ch/hb2010/papers/tho1d01.pdf

[27] Reiser M 2008 *Theory and Design of Charged Particle Beams* 2nd edn (Weinheim: Wiley-VCH)

[28] Dubin D H E and O'Neil T M 1999 Trapped non-neutral plasmas, liquids, and crystals (the thermal equilibrium states), *Rev. Mod. Phys.* **71** 87–172

[29] https://news.fnal.gov/2018/11/single-electron-beam-observed-in-iota-for-the-first-time/

[30] Sessler A M 1998 Gamma-ray colliders and muon colliders, *Phys. Today* **51** 48

[31] Symon K R, Kerst D W, Jones L W, Laslett L J and Terwilliger T M 1956 Fixed-field alternating-gradient particle accelerators, *Phys. Rev.* **103** 1837

[32] Craddock M K and Symon K R 2008 Cyclotrons and fixed-field alternating-gradient accelerators *Rev. Accel. Sci. Technol.* **1** 65–97

[33] Symon K R 2003 MURA days *Proceedings of the 2003 Particle Accelerator Conf. PAC03* (Piscataway, NJ: IEEE) pp 452–456

[34] *The Rebirth of the FFAG*, CERN Courier 27 July 2004 https://cerncourier.com/a/the-rebirth-of-the-ffag/

[35] Takahashi K *et al* 2012 Magnetic field design of coil-dominated magnets wound with coated conductors *IEEE Trans. Appl. Supercond.* **22** 4901705

[36] Craddock M K 2011 FFAG OPTICS, presentation at Int. Workshop on FFAG Accelerators (FFAG'11), Daresbury Laboratory, september

[37] Machida S 2013 Fixed field alternating gradient *CERN Accelerator School Proc. CERN 2013-1* https://cds.cern.ch/record/1514476?ln=en

[38] Johnstone C, Wan W and Garren A Fixed field circular accelerator designs *Proc. of the 1999 Particle Accelerator Conf. PAC99 (New York, NY)*

[39] Machida S *et al* 2012 Acceleration in the linear non-scaling fixed-field alternating-gradient accelerator EMMA *Nat. Phys.* **8** 243

[40] Muratori B 2011 Status of EMMA, presentation at Int. Workshop on FFAG Accelerators (FFAG'11), Daresbury Laboratory, September

[41] Barlow R *et al* 2010 EMMA—the world's first non-scaling FFAG, *Nucl. Instrum. Methods Phys. Res. A* **624** 1–19

[42] Machida S 2014 What We Learned from EMMA, presentation at Int. Workshop on FFAG Accelerators (FFAG'14), Brookhaven National Laboratory, September

[43] Okamoto H, Endo M, Fukushima K, Higaki H, Itoa K, Moriya K, Yamaguchi S and Lund S M 2014 Experimental simulation of beam propagation over long path lengths using radio-frequency and magnetic traps *Nucl. Instrum. Methods Phys. Res. A* **733** 119–28

[44] Gilson E P, Davidson R C, Efthimion P C and Majeski R 2004 Paul trap simulator experiment to model intense-beam propagation in alternating-gradient transport systems *Phys. Rev. Lett.* **92** 155002

[45] Gilson E P, Chung M, Davidson R C, Efthimion P C, Majeski R and Startsev E A 2005 Simulation of long-distance beam propagation in the Paul trap simulator experiment *Nucl. Instrum. Methods Phys. Res. A* **544** 171–8

[46] Okamoto H, Higaki H, Ito K, Kojima K, Matsuba M and Tokashiki Y 2021 Application of a compact ion trap to fundamental beam-dynamics studies, *Proc. 10th Int. Conf. Nuclear Physics at Storage Rings (STORI'17), JPS Conf. Proc.* vol 35 011007

[47] Martin L K *et al* 2019 A study of coherent and incoherent resonances in high intensity beams using a linear Paul trap *New J. Phys.* **21** 053023

[48] Davidson R C, Qin H and Shvets G 2000 A Paul trap configuration to simulate intense non-neutral beam propagation over large distances through a periodic focusing quadrupole magnetic field *Phys. Plasmas* **7** 1020

[49] Gilson E P, Davidson R C, Efthimion P C, Majeski R and Qin H 2003 Recent results from the Paul trap simulator experiment *Proc. of the 2003 Particle Accelerator Conf. PAC03 (Portland, OR)* (Piscataway, NJ: IEEE)

[50] Chung M, Gilson E P, Davidson R C, Efthimion P C and Majeski R 2009 Use of a linear Paul trap to study random noise-induced beam degradation in high-intensity accelerators *Phys. Rev. Lett.* **102** 145003

[51] Kojima K, Okamoto H and Tokashiki Y 2019 Empirical condition of betatron resonances with space charge *Phys. Rev. Accel. Beams* **22** 074201

[52] Hofmann I 2020 Comment on empirical condition of betatron resonances with space charge *Phys. Rev. Accel. Beams* **23** 028001

[53] Kojima K, Okamoto H and Tokashiki Y 2020 Reply to comment on empirical condition of betatron resonances with space charge *Phys. Rev. Accel. Beams* **23** 028002

[54] Hofmann I and Boine-Frankenheim O 2015 Space-charge structural instabilities and resonances in high-intensity beams *Phys. Rev. Lett.* **115** 204802

[55] Ito K, Okamoto H, Tokashiki Y and Fukushima K 2017 Coherent resonance stop bands in alternating gradient beam transport, *Phys. Rev. Accel. Beams* **20** 064201

[56] Takeuchi H, Fukushima K, Ito K, Moriya K, Okamoto H and Sugimoto H 2012 Experimental study of resonance crossing with a Paul trap, *Phys. Rev. ST Accel. Beams* **15** 074201

[57] Moriya K, Fukushima K, Ito K, Okano T, Okamoto H, Sheehy S L, Kelliher D J, Machida S and Prior C R 2015 Experimental study of integer resonance crossing in a non-scaling fixed field alternating gradient accelerator with a Paul ion trap *Phys. Rev. ST Accel. Beams* **18** 034001

[58] Okamoto H, Fukushima K, Higaki H, Ishikawa D, Ito K, Iwai T, Moriya K, Okano T, Osaki K and Yamaguchi M 2014 Beam dynamics studies with non-neutral plasma traps *Proc. 5th Int. Particle Accelerator Conf. IPAC (Dresden, Germany)* 4052 https://accelconf. web.cern.ch/ipac2014/papers/frxaa01.pdf

[59] Kelliher D J, Machida S, Prior C R, Sheehy S L, Fukushima K, Ito K, Moriya K, Okamoto H and Okano T 2014 Study of resonance crossing in non-scaling FFAGs using the S-POD linear Paul trap *Proc. 5th Int. Particle Accelerator Conf. IPAC (Dresden, Germany)* 1571 https://accelconf.web.cern.ch/ipac2014/papers/tupri009.pdf

[60] Fukushima K and Okamoto H 2015 Design study of a multipole ion trap for beam physics applications *Plasma Fusion Res.* **10** 1401081

[61] Martin L K *et al* 2019 Can a Paul Ion Trap Be Used to Investigate Nonlinear Quasi-Integrable Optics? *J. Phys. Conf. Ser.* **1350** 012132

[62] Martin L 2019 Progress towards testing non-linear integrable optics in a Paul trap, presentation at the 4th ICFA Mini Workshop on Space Charge, CERN https://cds.cern.ch/record/2699737?ln=en

IOP Publishing

A Practical Introduction to Beam Physics and Particle Accelerators (Third Edition)

Santiago Bernal

Appendix A

Computer resources and their use

Instead of having end-of-chapter problems and exercises, this book includes a 'Computer Resources' section in each chapter. Therefore, exploration and the pace of learning do not depend on completing a minimum number of problems. Instead, the reader is free to explore the available resources, and only he or she can decide when the math, the physics, and the software all come together to the point of 'understanding,' before moving on to the next chapter.

Computer software has been an integral part of learning any technical or scientific field; this has particularly been the case since the popularization of personal computers more than forty years ago. We can use software of three general types: low-level and high-level programming codes, and specific application software. We do not cover here (except for a very brief description of Python) the use of popular languages such as C++, Python, or Java. Python, in particular, is very popular for writing input files and scripts in several particle accelerator codes. The names of some codes (e.g., *JMAD* and *PyOrbit*) reveal their dependence on Java or Python. Furthermore, Java is the basis for many 'applets' available online as educational tools in many areas. (Unfortunately, Java applets cannot be run online anymore, at least with the best-known browsers.)

In this book, we advocate the use of *Mathcad* and *Matlab* or their freeware counterparts *SMath Studio* and *Octave* for general calculations and programming, with no prejudice towards the powerful *Mathematica,* which we think is superior to the rest for symbolic calculations. In addition to *Mathematica*, Wolfram Research has created a stand-alone educational tool called 'CDF Player' that is essentially a run-only *Mathematica*. A large number of CDF demonstration projects are freely available from Wolfram and third parties. As for specific application software, we concentrate here on a number of popular codes for particle accelerator design and on a few general optics programs (mostly for chapters 1 and 2).

All the particle accelerator codes mentioned below, except for *PBO Lab* and some auxiliary software such as *VMware Fusion*, are freely available and can be installed on the most popular operating systems without any special difficulty. With the exception of *MARYLIE* and *WARP*, no code compilation is required, but it may be necessary to run scripts to build packages or implement the code in a virtual machine or emulation software (e.g., Wine/PlayOnLinux under Ubuntu Linux, or the new *Ubuntu app* under Windows 10). Furthermore, a few apps for Google's Android and Apple's IOS will be mentioned because of their instructional value.

In our opinion, the best approach to learning any of the codes is by example, and not by reading manuals. Manuals should be used only as references and initially to check for general information such as sign conventions and important approximations (e.g., relativistic vs. non-relativistic, symplectic vs. non-symplectic, zero-current vs. space-charge effects, etc).

The following steps may be a good guide to learning not only the codes but also the physics:

Open one of the example files provided and run the code to familiarize yourself with the basic graphical user interface (GUI) (or command-line) mechanics and capabilities; (2) edit the file to explore the scaling of different parameters; (3) do 'sanity' checks, i.e., back-of-the-envelope or Mathcad calculations to make sure the code is yielding the expected results; (4) create your own concept design and either modify an existing file or start from scratch to implement your model in the code; (5) go back to (3); and (6) try a different code for cross-checking, or simply to compare programs and decide on which one to use. Regarding the last point, it is advisable to learn to be reasonably proficient in one code before switching to a different one. Naturally, there is no single code that can do everything that you will eventually need, so learning a few codes may be a good idea.

Tables A.1–A.3 summarize the computer codes that are described in the following sections. Not all compatibility issues have been explored in depth, so it is possible that some of the codes can actually run in a particular operating system (OS) despite the annotation 'x' or a blank cell. The book's website provides example files for most codes https://ireap.umd.edu/clark/faculty/1273/Santiago-Bernal.

A.1 Hamiltonian dynamics and the symplectic condition

As discussed in the first section of chapter 4, Hamilton's equations of motion in classical mechanics lead to Liouville's theorem. This theorem asserts that the flow of representative points in phase space for systems without dissipation is similar to the flow of an incompressible fluid. Equivalently, the evolution of Hamiltonian systems can be described in terms of *canonical transformations,* which can be represented by *symplectic* matrices. A matrix M is symplectic if it satisfies the condition [1]:

$$MJM^T = J, \quad \text{or} \quad M^TJM = J \tag{A.1}$$

where J is the antisymmetric matrix defined by

$$J = \begin{bmatrix} 0 & 1 \\ -1 & 0 \end{bmatrix}. \tag{A.2}$$

Table A.1. Code compatibility for the particle accelerator programs used in the book.

Computer code ↓	Operating systems					
	Microsoft		Linux		Apple	
	Win 10[1]	Win 10+ Cygwin	Ubuntu 20.04 LTS	Ubuntu 20.04 LTS + Wine	MAC OSX	MacOS+ VMware Fusion[4] or Wine[5]
MAD-8	✓		✓			
MAD-X	✓		✓		✓	
ELEGANT	×	✓	✓	×	✓	✓
WINAGILE	×	×	×	✓	×	✓
MARYLIE	×	×	✓	×	✓	✓
SPOT[2]	×	×	×	✓	×	✓
MENV[3]	✓	×	✓	×	✓	✓
TRACE 2-D	✓	×	×	✓	×	✓
TRACE 3-D	✓	×	×	✓	×	✓
WARP	×	×	✓	×	✓	

[1] Ubuntu app available
[2] Originally written for Win 3.1
[3] MatLab program
[4] VMware Fusion + Ubuntu Linux or + Windows XP
[5] Wine (PlayOnLinux) under Ubuntu Linux, or MacOS Wine.

Table A.2. Code compatibility for the optics software and other packages used in the book.

Computer code or App ↓	Operating systems					
	Microsoft		Linux		Apple	
	Win 10	Win 10+ Cygwin	Ubuntu 20.04 LTS	Ubuntu 20.04 LTS + Wine	MAC OSX	Mac OSX+ VMware Fusion
Optgeo	✓	×	✓	×	✓	✓
2D Ray Tracer[1]	✓	×	×	×	×	✓
Ray Optics				Android App		
RayLab				IOS App		
TAPAS				Android App		
Radiation2D	✓	×	×	✓	×	✓
CDF Player	✓	×	✓	×	✓	✓

[1] Mathcad program.

For a system with n degrees of freedom, the symbol 0 stands for an $n{\times}n$ matrix composed of zeros as elements, and 1 is an $n{\times}n$ unit matrix. In Hamiltonian mechanics M represents the $2n{\times}2n$ *Jacobian* matrix of a canonical transformation. Using this language, Hamilton's equations of motion for a system of n degrees of freedom are written as

Table A.3. Computer code homepage or relevant website.

Computer code ↓	Homepage / Websites
MAD-8	http://project-madwindows.web.cern.ch/MAD-8/default.htm
MAD-X	http://mad.web.cern.ch/mad/
ELEGANT	https://www.aps.anl.gov/Accelerator-Operations-Physics/Software
WINAGILE	https://www.terpconnect.umd.edu/~sabern/PIBPPA-3rdEd/Appendix%20A/
MARYLIE	http://www.physics.umd.edu/dsat/ http://www.ghga.com/accelsoft/
SPOT[1]	https://terpconnect.umd.edu/~sabern/PIBPPA/Chapter%206/SPOT/
MENV[2]	https://terpconnect.umd.edu/~sabern/PIBPPA/Chapter%206/MenV/
TRACE 2-D	http://laacg.lanl.gov/laacg/services/download_trace.phtml
TRACE 3-D	http://laacg.lanl.gov/laacg/services/download_trace.phtml
WARP	http://warp.lbl.gov/
Optgeo	http://jeanmarie.biansan.free.fr/optgeo.html
Ray Optics	https://play.google.com/store/apps/details?id=com.shakti.rayoptics&hl=en
2D Ray Tracer	http://kodu.ut.ee/~kiisk/mcadapps/ Look for 'raytrace.zip'
RayLab	https://itunes.apple.com/us/app/raylab/id710190065?mt=8
TAPAS	https://play.google.com/store/apps/details?id=borland.TAPAs&hl=en
Radiation2D	https://groups.oist.jp/qwmu/software
CDF Player	http://www.wolfram.com/cdf-player/

$$\dot{\eta} = J\frac{\partial H}{\partial \eta}, \tag{A.3}$$

where η represents a column vector with $2n$ elements: $\eta_i = q_i$, $\eta_{i+n} = p_i$ for $i \leqslant n$. H is the Hamiltonian, and the dot indicates the time derivative. Equation (A.3) is a shorthand notation for $2n$, first-order differential equations with a sign asymmetry between the 'p' and 'q' equations.

Numerical methods involve transforming the initial state of the system over many successive time intervals. In general, however, these transformations are not symplectic and thus do not represent the evolution of a Hamiltonian system. For example, the simple first-order implicit or explicit numerical Euler methods (see any textbook on numerical methods) are not symplectic; even the standard fourth-order Runge–Kutta method, the workhorse of numerical work, is not symplectic. Therefore, the use of these non-symplectic 'integrators' often yields unphysical results, particularly for simulations over extended integration intervals (e.g., instabilities not seen in real accelerator experiments). Naturally, this is not acceptable for studies involving, for example, hundreds of thousands or millions of turns in circular accelerators, or simulations of the evolution of galactic systems over millions of years. Furthermore, in matrix codes such as TRANSPORT [2], the initial particle coordinates are transformed by first- and second-order matrices through most lattice elements (see equation (1.15)); this amounts to a truncation of the Taylor-series expansion of the Hamiltonian, which renders the transformation non-symplectic.

It is possible, however, to construct *symplectic integrators*. Numerical symplecticity not only yields the correct Liouvillian flow, but also prevents the numerical total energy from deviating monotonically from the correct conserved value. For example, a combination of implicit and explicit first-order Euler methods yields a first-order symplectic integrator. Dennis Donnelly and Edwin Rogers discuss this case and other higher-order methods in the context of the simple harmonic oscillator in an *American Journal of Physics* article [3]. More general discussions of symplectic integrators can be found in an article by H Yoshida in the *Handbook of Accelerator Physics and Engineering* [4], and the books by Étienne Forest [5], or Andrzej Wolski [6].

Tracking through *thin* elements is symplectic; we can use the matrix for a thin lens, equation (1.5), and equation (A.1) above to prove it. Although symplectic tracking using *thick* elements is also possible, the computational speed is greater when thin elements are used. We have included a simple *Mathcad* document, **Simplecticity.xmcd**, in the book's website, to test the symplecticity of thin and thick lens matrices.

The MAD-8 and MAD-X codes (section A.5.1) have the command 'MAKETHIN,' which turns thick elements into symplectic thin ones. In the ELEGANT code (section A.5.2), non-symplectic sector magnets (SBEN) or quadrupoles (QUAD), or their symplectic counterparts CSBEND and KQUAD can be used. The canonical bend CSBEND, for example, is described in the ELEGANT manual as using the *exact* Hamiltonian, i.e., all orders of the momentum offset and curvature. This is particularly important for tracking in rings.

Non-Liouvillian processes, such as the generation of synchrotron radiation, are non-symplectic. Even in this case, tracking is best performed with symplectic integrators while ignoring radiation. However, the use of all approximations, especially the *paraxial approximation*, must be checked, because it may yield larger errors than those originating from non-symplectic elements. This is particularly important for tracking in small rings.

Last, but very significantly, it is possible to preserve simplecticity with calculations involving a truncated Taylor-series expansion (in canonical variables) of the Hamiltonian. The single-particle code *MARYLIE*, described briefly in chapter 1 and in section A.5.4, is a primary example of a symplectic map code. It utilizes an approach to Hamiltonian mechanics based on Lie algebra methods which are not discussed in most accelerator physics textbooks (Wolski's book [6] is an exception). We give below a summary of the main ideas behind this approach. The reader is encouraged to read also an excellent overview by Dragt [7], chapter 3 of the *MARYLIE* manual [8], and/or consult the detailed monograph in [9]. For a comprehensive, almost encyclopedic discussion, see [10] (available online).

A.2 Lie algebra methods

Hamilton's equations (A.3) can be recast in terms of *Poisson brackets*. If u and v are two functions of the canonical variables contained in η, the Poisson bracket of u and v is [1]:

$$[u, v]_\eta = \left(\frac{\partial u}{\partial \eta}\right)^T J \frac{\partial v}{\partial \eta}.$$ (A.4)

Thus, if $v = H$, the Hamiltonian, and u is η_i, then

$$[\eta, H]_\eta = J \frac{\partial H}{\partial \eta} = \dot{\eta},$$ (A.5)

from equation (A.3). In Dragt's notation, equations (A.4) and (A.5) read

$$[u, v]_\eta = : u : v,$$ (A.6)

$$\dot{\eta} = - : H : \eta.$$ (A.7)

Now, if $\boldsymbol{\eta}^{in}$ and $\boldsymbol{\eta}^{fin}$ represent the vectors of the *initial* and *final* canonical coordinates, respectively, of a particle as it moves through a transport line, then

$$\boldsymbol{\eta}^{fin} = M\boldsymbol{\eta}^{in},$$ (A.8)

where M stands for a symplectic *transfer map*. The form of equation (A.7) suggests that the solution to equation (A.8) can be written as $\eta(t) = e^{-t:H:}\eta(0)$, or, equivalently, $M = e^{-t:H:}$, where the exponential function is defined by the standard Taylor-series expansion. In general, the map can be written as an infinite product of *Lie transformations* (factorization theorem):

$$M = e^{:f_1:}e^{:f_2:}e^{:f_3:}e^{:f_4:} ... = e^{:f_1:}Re^{:f_3:}e^{:f_4:} ... ,$$ (A.9)

where f_m are homogeneous polynomials of degree m. The Lie transformation $e^{:f_1:}$ describes the effects of misalignment or power errors in the beamline. The *linear* part of the map is $R = e^{:f_2:}$, and $e^{:f_3:}$, $e^{:f_4:}$, etc. represent the second-, third-, and higher-order effects. As seen above, the explicit form of M depends on the Hamiltonian; for a one-dimensional simple harmonic oscillator, for example, $H = (1/2)\left[p_x^2 + \omega^2 x^2\right]$, so $f_2 = -Ht$, leading to the standard solution after repeated application of Poisson bracket relations such as $:H: x = -p_x$, $:H:^2 x = -\omega^2 x$, etc [6, 11].

A.3 Software and hardware

This book and the programs developed for it were written using software for Microsoft Windows 10. Some programs were also tested using MacOS 11.6.5 (Big Sur) and Ubuntu Linux 20.04 LTS; the latter was used with and without the Windows emulation application *Wine* (PlayOnLinux), which is also available for MacOS. *Wine* was particularly useful for running old software written for Microsoft Windows XP or 3.1 (see the tables above). A new Ubuntu app for Windows 10, which requires some effort to install and is essentially an Ubuntu Linux terminal, let us compile and run FORTRAN and C code using standard Linux compilers. In addition, VMware Player and VMware Fusion were employed to implement virtual machines. We also tested some applications on iOS and Android mobile devices.

A.4 General tools

A.4.1 Mathcad

Mathcad is a self-documenting software intended for general calculations in the fields of math, science, and engineering. The code is particularly appealing for those unwilling to learn a programming language who are interested in applications in which speed is not the main factor. *Mathcad*, introduced in 1986 for DOS by MathSoft, is now owned by PTC. PTC Mathcad is currently only available for Microsoft Windows, but can be run in MacOS or Linux through a Windows virtual machine. PTC announced 'end-of-sale' for *Mathcad 15* and *Prime 1.0–6.0* in December 2021; the only versions available for subscription at the time of writing (April 2022) are *Prime 7.0* and the current *Prime 8.0*. A one-year academic license is available for $60; others can obtain a free 30 day trial version that will revert to *PTC Mathcad Express* if the full license is not purchased by the end of the trial period. *PTC Mathcad Express* will retain only the basic features of *PTC Mathcad*. There is also a freeware alternative to *PTC Mathcad* called *SMath Studio* (or *SMath Solver*) written by Andrey Ivashov, which is available for Windows, Linux, Android, and iOS. *SMath Studio* does not have all the functionality of *Mathcad*, but it is getting better with every new release.

Almost all the numerical examples quoted in the 'Computer Resources' sections of the book were written in *Mathcad 14* and *15*, and all the related files are available on the book's website, which includes some SMath Studio versions as well. In addition, we are uploading *Prime 8.0* versions of the files as they are developed (*Prime 8.0* has a convenient conversion utility). In figure A.1 we show a *Mathcad* document for the solution of the damped harmonic oscillator. We present equivalent examples in *Matlab* and Python below.

A.4.2 Matlab

Matlab is possibly the most popular general computing tool for engineers, scientists, and educators worldwide. It was initially developed in the 1970s and 1980s by Cleve Moler and others from the University of New Mexico and Stanford University. Matlab's name derives from 'Matrix Laboratory,' as its original signature strength was, and still is, matrix calculations. *Matlab* is owned by MathWorks and is available for all major platforms. Several freeware *Matlab* 'clones' have appeared over the years, most notably *Octave* and *Scilab*, for various operating systems. Although not as transparent as *Mathcad, Matlab's* learning curve is well within the abilities of freshmen, and, as with many programming languages, can be learned faster using existing sample programs or templates. In fact, many free as well as commercial packages or 'toolboxes' for various applications (e.g., signal processing, image processing, data acquisition and control, etc.) are available, including one for particle accelerators. The latter is called the 'Accelerator Toolbox' or just 'AT' and was introduced by Andrei Terebilo from Stanford Linear Accelerator Center (SLAC) in 2001.

To illustrate the use of *Matlab*, we present below a simple example and instructions for solving the damped harmonic oscillator differential equation. (Obviously, there is an analytical solution to this problem, and multiple ways within *Matlab* of finding numerical solutions.)

DAMPED SIMPLE HARMONIC OSCILLATOR

5-29-2015

$m := 1$ — Mass in kg

$k := 100$ — Spring constant in N/m

$b := 1$ — Damping constant in Ns/m

$IC := \begin{pmatrix} 1 \\ 0 \end{pmatrix}$ — Initial conditions, $x(0) = 1$ m, $x'(0) = 0$ m/s

$D(t,x) := \begin{pmatrix} x_1 \\ \dfrac{-k}{m} \cdot x_0 - \dfrac{b}{m} \cdot x_1 \end{pmatrix}$ — Differential equation vector. First element is dx/dt, and second element is $d^2x/dt2$.

$M := 1000$ — Integration points

$Sol := rkfixed\,(IC, 0, 5, M, D)$ — Solution array using 4th order Runge-Kutta method with fixed step. First column (0) is time, second column (1) is "x", and third column (2) is dx/dt.

$i := 0 .. M$

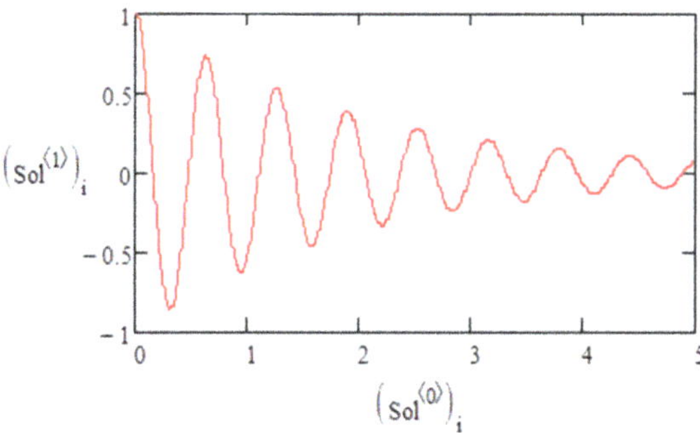

Figure A.1. Simple example of a Mathcad document.

Write *and* save the '*M-file*' named `oscillator.m` (use the built-in script editor):

```
function xpp = oscillator(t,x)
m = 1; k = 100; b = 1; % mass, spring and damping constants (SI units)
xpp = zeros(2,1);
xpp(1) = x(2);
xpp(2) = -(k/m)*x(1)-(b/m)*x(2);
end
```

In the *command* window, run

```
[t,x]=ode45(@oscillator, [0 5], [1 0]); plot(t,x(:,1))
```

to generate *x* vs. *t*. Compare it with the Mathcad document in figure A.1 above.

We have translated some of this book's *Mathcad* documents into *Matlab* (see this book's website).

A.4.3 Python

Python is an extremely popular general-purpose programming language. Python's syntax is not unlike Matlab's, and their speeds are comparable, as both are interpreted languages. Python's trademarks are clarity and simplicity; it is highly readable and extensible, with a large body of numerical and scientific libraries or packages. Python is freely available for all major platforms. Some implementations, such as *Anaconda*, come with an excellent editor (*Spyder*) and all major scientific packages (e.g., numpy, matplotlib, simpy). Guido van Rossum from the Netherlands created Python in 1989. Python 2.0 was released in 2000, and Python 3.0 in 2008. Python 3.8 is installed by default in Ubuntu Linux 20.04, but Python 2.7 (April 2020) can be installed as well. With these versions, it is possible to run any one of thousands of programs generally available in websites associated with computer and physics courses at universities worldwide. Using these programs is also a good way to learn Python by example. In addition, special versions of Python are available: *IPython*, which runs 'by cells' similar to those of *Mathematica*, and a visual Python that facilitates the graphical simulation of simple and some complex physical systems.

To conclude this section, we show a simple Python program to solve the damped simple harmonic oscillator problem. (The module 'scipy' must be installed first.)

A key difference between Python and other languages is that *indenting* is crucial in Python. Compare the Python script below with *Matlab*'s version. Note that in Python, unlike *Matlab*, there is no need to complete two operations to run the code (the *M-file* plus the command instruction). Another important difference is that indexing of arrays starts at zero in Python and one in *Matlab*. In Mathcad, the start index is zero by default, but can be changed to one by typing ORIGIN:=1.

```python
"""Damped Simple Harmonic Oscillator S. Bernal, 5-29-2015"""
from scipy.integrate import odeint
from pylab import *   # for plotting commands

def deriv(x,t): # return derivatives of the array x
    m = 1    # Mass in kg
    k = 100   # spring constant in N/m
    b = 1     # damping constant in Ns/m
    return array([ x[1],-(k/m)*x[0]-(b/m)*x[1] ])

time = linspace(0.0,5.0,1000)
xinit = array([1.0,0.0]) # initial values
x = odeint(deriv,xinit,time)

figure()
plot(time,x[:,0],'r-') # x[:,0] is the first column of x
xlabel('t')
ylabel('x')
show()
```

Other general software can be employed to model and perform design calculations for accelerators. For example, Microsoft Excel has been used by David Douglas *et al* at Jefferson Lab in Virginia [12] to model the basic linear optics of the Free-Electron Laser (FEL) driver.

A.5 Matrix/map codes

A.5.1 MAD-8 and MAD-X

The Methodical Accelerator Design (MAD) program has been used at CERN for over 20 years to simulate, design, and optimize large linear and circular particle accelerators. A few major examples are the Proton Synchrotron (PS), the Super Proton Synchrotron (SPS), the Large Electron–Positron (LEP) collider, and the Large Hadron Collider (LHC). The old version, MAD-8, was 'retired' in 2002, but at least two versions are still available for download (see table A.3): 8.23.06 and 8.51/15 from SLAC and CERN.

MAD does not have a GUI, as it is more a *language* than a code. Its User's Reference Manual [13] contains useful physics discussions and detailed explanation of all the commands, syntax, etc. but very little on how to run the code. On the other hand, there are a few very useful tutorial presentations available online. We have adapted an example from one of them [14] for use with the latest (October 2017) MAD-X release, version 5.03.07. We emphasize that we are only scratching the surface of a very powerful and complex program for single-particle simulations of accelerators. Space-charge effects can also be simulated, but in conjunction with another code called Polymorphic Tracking Code (PTC). The CERN website has more details.

DOWNLOAD: To download MAD-8 or MAD-X, use the hyperlinks in table A.3. For MAD-X, click 'Releases' on the left list and then 'releases repository.'

INSTALLATION: for Windows 10, move the madx-win64.exe (64-bit Win) or madx-win32.exe (32-bit Win) file to a folder 'MAD-X'. Rename the file to simply 'madx.exe.'

RUN / EXAMPLE: the MAD-X language is case insensitive. The file below, FODOUMERrev.MADX, is self-explanatory for the most part. A focusing–defocusing (FODO) unit cell for the University of Maryland Electron Ring (UMER) machine is defined, and the beta functions and phase advances are plotted for arbitrary initial values BETX=0.9, BETY=0.7, ALFX=0, ALFY=0. The calculations are repeated for *periodic* matched conditions ('cell matching' in the terminology of the MAD User's Reference Manual) and for assumed phase advances of 90°. Comments are preceded by //, /*, or !. Note that the energy 'PC' is in units of GeV/c (very small for the 10 keV UMER!)

```
//FILE: FODOUMERrev.MADX
//MADX Example 1: FODO cell
//Author: L. Deniau (CERN), S. Bernal (U. of MD)
//Date: July 6th, 2020
```

```
BEAM, PARTICLE=ELECTRON, PC=10158E-06;
D : DRIFT, L=0.10836;
QF: QUADRUPOLE, L=0.05164/2, K1= 168.36;
QD: QUADRUPOLE, L=0.05164, K1=-168.36;
FODO: LINE=(QF,D,QD,D,QF);
USE, PERIOD=FODO;
! Non periodic solution
TWISS, BETX=0.9, BETY=0.7, ALFX=0, ALFY=0;
MYPLT: PLOT, HAXIS=S, HMAX=0.35,
           VAXIS1={BETX,BETY}, VAXIS2={MUX,MUY},
           COLOUR=100, INTERPOLATE;
! Periodic solution
TWISS;
EXEC, MYPLT;
! Match phase advance to 90 degrees
MATCH, SEQUENCE=FODO;
VARY, NAME=QF->K1;
VARY, NAME=QD->K1;
CONSTRAINT, RANGE=#E, MUX=0.25;
CONSTRAINT, RANGE=#E, MUY=0.25;
LMDIF, TOLERANCE=1e-20;
ENDMATCH;
! Matched solution
TWISS;
EXEC, MYPLT;
```

The program is run by typing

```
madx FODOUMERrev.MADX
```

The output appears in the form of the postscript file 'madx.ps,' which can be opened with GhostView in Windows. In the example above, three plots are generated (one per page). The first two plots are shown in figure A.2. Additional examples can be found in [15].

A.5.2 Elegant

ELEGANT is a powerful third-order matrix code widely used to model light sources and other accelerators, from the Advanced Photon Source (APS at Argonne National Lab) to X-ray FELs such as the Linac Coherent Light Source (LCLS), at Stanford National Lab. ELEGANT was created by Michael Borland from APS in 1988 and is maintained by him and others at APS and elsewhere [16, 17]. There is

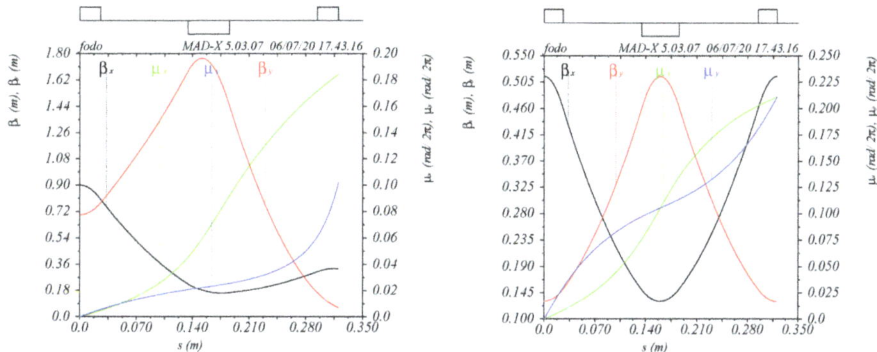

Figure A.2. Output plots from MAD-X: unmatched FODO cell (left), and matched FODO cell (right). Notice the diagram of the FODO cell above each plot.

also an ELEGANT forum, where users can almost always find solutions to code or model-related problems. The code is freely available from the APS website for all major platforms, but you will need to register with your e-mail to get a key to download the files and documents (ELEGANT, SDDS, examples, and manuals).

DOWNLOAD: You can download all the files from the APS website. See table A.3.

INSTALLATION: For the Windows installation, one should first install CYGWIN and Microsoft Visual C++. Follow the steps provided by the APS website. Installation in Ubuntu Linux is relatively straightforward; we recommend building the 'rpm' packages using the script in Step 1 instead of trying the pre-built RPMs. For MacOS, download the binary executables for both Self-Describing Data Sets (SDDS) and ELEGANT (darwin-x86 compressed folders); ELEGANT will run on the latest MacOS (12.3 Monterey—at the time of writing, May 2022) in machines with the new M1 chip or Intel chips.

RUN / EXAMPLE: the main files in ELEGANT are the *input* file, which has an '.ele' extension, and the *lattice* file, which has an '.lte' extension. These files can be edited with any text editor, such as WordPad in Windows, 'gedit' in Linux, or TextEdit on the Mac. Shown below is an example of a *lattice file*, 'umer1.lte', corresponding to the UMER:

```
!umer1.lte
!Simple Example from UMER
DR1: DRIF,L=0.05418
QD1: QUAD,L=0.05164,K1=-153.419
DR2: DRIF,L=0.03538
DIP1:  SBEND,L=0.0376,ANGLE=0.1745329252,E1=0.0872664626,
E2=0.0872664626,HGAP=0.0287,FINT=0.3,&
FSE=-0.0
DR3: DRIF,L=0.03538
QF1: QUAD,L=0.05164,K1=153.419
DR4: DRIF,L=0.05418
```

```
DR5: DRIF,L=0.05418
QD2: QUAD,L=0.05164,K1=-153.419
DR6: DRIF,L=0.03538
DIP2:  SBEND,L=0.0376,ANGLE=0.1745329252,E1=0.0872664626,
E2=0.0872664626,HGAP=0.0287,FINT=0.3,&
FSE=-0.0
DR7: DRIF,L=0.03538
QF2: QUAD,L=0.05164,K1=153.419
DR8: DRIF,L=0.05418
!Define Beamlines
FODO1: LINE=(DR1,QD1,DR2,DIP1,DR3,QF1,DR4)
FODO2: LINE=(DR5,QD2,DR6,DIP2,DR7,QF2,DR8)
SECTION: LINE=(FODO1,FODO2)
RING: LINE=(18*SECTION)
```

The notation and structure of this file are similar to those of the files used for the MAD codes (section A.5.1). UMER consists of 18 sections, each of which has two FODO periods. There are two bend dipoles and four quadrupoles per section (see section 9.1 for additional details). 'K1' represents the quadrupole focusing constant κ_0 in m^{-2}; the bend dipoles are rectangular 10° bend magnets (two 5° edge angles, E1 and E2), and FINT is a constant ('field integral') that is introduced to implement edge effects (see the TRANSPORT manual [2] for its definition). Note that '!' must precede comments. The ELEGANT manual [17] contains a detailed description of a great number of accelerator lattice elements and their parameters.

The ELEGANT input file is 'umer1.ele':

The Elegant input file is 'umer1.ele':

```
!umer1.ele

  !Simple Example from UMER

  &divide_elements
     maximum_length=0.002
     name = *
  &end

  &run_setup
    lattice = umer1.lte
    default_order = 3
    rootname = umer1
    use_beamline = "ring"
    output = %s.out
```

```
  sigma = %s.sig
  p_central = 0.199
  final = %s.fin
  magnets = %s.mag
&end

&run_control
  n_steps = 1
  n_passes=5
&end

&twiss_output
   matched = 1,
   beta_x = 3.119e-1,
   alpha_x = 1.324,
   beta_y = 3.090e-1,
   alpha_y = -1.315,
   eta_x = 0.0,
   etap_x = 0.0,
   eta_y = 0,
   etap_y = 0,
   filename = %s.twi
&end

!&bunched_beam
   bunch = %s.bun
   n_particles_per_bunch = 10000,
   emit_x = 8.048e-6,
   emit_y = 8.048e-6,
   beta_x = 3.119e-1,
   alpha_x = 1.324,
   beta_y = 3.090e-1,
   alpha_y = -1.315,
   eta_x = 0.0,
   etap_x = 0.0,
   eta_y = 0,
   etap_y = 0,
   use_twiss_command_values = 0,
   Po = 0.199,
   sigma_dp = 0.01,
   sigma_s = 1.69,
   distribution_cutoff[0] = 2.5
   distribution_cutoff[1] = 2.5
   distribution_cutoff[2] = 1.73
```

```
            distribution_type[0] = "gaussian"
            distribution_type[1] = "gaussian"
            distribution_type[2] = "hard-edge"
            save_initial_coordinates =1;
        !&end


        !&track &end

        &save_lattice filename = %s.new
            output_seq = 1
        &end

        &stop &end
```

Note that the 'umer1.ele' file shown above is divided into blocks that begin with '&name of block' and end in '&end'. For example, we have '&run_setup,' '&twiss_output,' '&run_control,' '&bunched_beam,' '&track,' and others. The elements of all available blocks are explained in detail in the ELEGANT manual. Furthermore, note how the lattice file and the beamline are called in the '&run_setup' block. Also, an entire block can be commented out using '!', as is the case with the '&bunched_beam' block above.

The code is run by typing

elegant umer1.ele

In the example above, the code finds the matched lattice functions Betax, Betay, Etax, and Etay, i.e., the transverse betatron and dispersion functions. To plot Betax, for example, type

sddsplot umer1.twi -col=s,betax

Notice the command 'sddsplot'; it is part of the SDDS tools, an important component of ELEGANT. To print the betatron tunes, one would type

sddsprintout umer1.twi -par={nux,nuy}

In order to track and plot *beam* functions, as opposed to *lattice* functions, we need the '&bunched_beam' block in the '.ele' file. Thus, we specify the initial beam parameters, distribution types in x, y, and z, and the number of particles (10 000 in the example). As an example, we can type

sddsplot umer1.sig -col=s,Sy

to plot the rms envelope in the vertical plane as a function of the reference-orbit coordinate 's'. Figure A.3 shows output plots of Betax produced using `use_beamline = "ring"` and `use_beamline = "section"` in the '&run_setup' block.

Running ELEGANT, plotting lattice and beam functions, and performing complex operations can be made very convenient by writing SDDS scripts. For example, we can have a simple script, 'simple_script.txt' that runs 'umer1.ele'; displays the betatron tunes, horizontal and vertical chromaticities, and momentum compaction; and plots Betax, Betay:

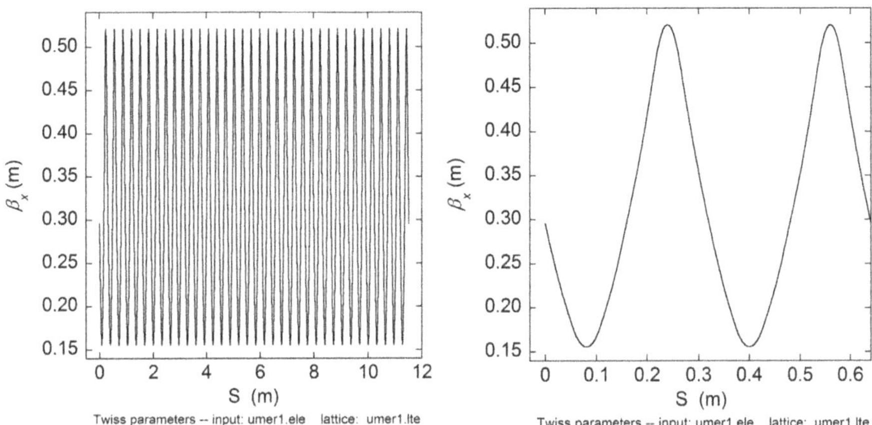

Figure A.3. Horizontal betatron functions calculated in ELEGANT: (left) one turn or 18 sections, and (right) one section of the UMER.

```
sddsmakedataset settings.sdds \
elegant umer1.ele
sddsprintout umer1.twi -par={nux,nuy}
sddsprintout umer1.twi -par={dnux/dp,dnuy/dp}
sddsprintout umer1.twi -par=alphac
sddsplot umer1.twi -col=s,betax
sddsplot umer1.twi -col=s,betay
```

The script can be run by typing './simple_script' (without quotation marks) in Microsoft Windows (under Cygwin), but the text file must be first converted to a Unix text file using the command 'd2u simple_script.txt'; 'd2u' stands for 'DOS-to-Unix.' This conversion is not required in Linux or MacOS. Finally, a very useful ELEGANT command is 'sddsquery filename'; for example, 'sddsquery umer1.twi,' 'sddsquery umer1.sig' lists all parameters that are part of the lattice and beam sets, respectively.

A.5.3 Winagile

WINAGILE is a free, (old) Microsoft Windows-based, computer code for beam transport calculations in transfer lines and rings. It can be used for particle tracking, beam envelopes, closed orbits and corrections, resonance studies, and many other applications [18]. Both incoherent and coherent space-charge effects can be included as well, but its models are not very accurate. The code was written by P. J. Bryant from CERN, who is also the author, with Kjell Johnsen, of *The Principles of Circular Accelerators and Storage Rings* [19]. With some guidance, one can learn how to run the basic WINAGILE in a matter of minutes. The GUI is not the best in the world, plus more than a few bugs remain in the available version (ver. 3.5), but it is nonetheless a working code which can be easily applied to model many accelerators.

DOWNLOAD: One can download the code from this book's website (see table A.3). The file 'package(ver3.5_D054).zip' includes the windows-executable file, lattice examples (extension 'LAT'), and the help file 'winagile.hlp.'

INSTALLATION/RUNNING: the code runs under Microsoft Windows XP, or under *Wine (PlayOnLinux)* in Ubuntu Linux. It is possible to run WINAGILE in Windows 10 but only inside a virtual Windows XP machine through VMware Player or Oracle's VirtualBox.

EXAMPLE: thinl1l2.lat—this lattice has two thin quadrupoles of opposite strengths that are separated by a distance shorter than their individual focal lengths, $f = 0.1$ m (magnitude), (see chapter 2). The total length of the beamline is 1.0 m.

(1) Run WINAGILE and open the example file **thinl1l2.lat** (which can be found in this book's website). This lattice contains two types of element: **DRIFT** and **THINQ.**

(2) Note that you are in the **MAIN WINDOW.** Explore the different column values. Only the 'Length' and 'Thin quad kL' columns have values. In particular, the length of THINQ is zero, by definition. The integrated strengths (in m^{-1}) of the quadrupoles have opposite signs.

(3) In the 'Calculations' menu choose 'Transfer Line.' A window 'Twiss Input' opens; click OK. The input here is irrelevant to our present calculations. You end up with a new window called **BEAM PARAMETERS.**

(4) In the 'Calculations' menu choose 'Tracking'. Enter 0.005 for 'Horizontal position, x' and also for 'Vertical position, z'. The window now changes to **TRACKING.**

(5) In the 'Graphs' menu choose 'Trajectories'. A window with a blue background opens showing the horizontal and vertical trajectories. The default particles are 1 GeV protons, but this is irrelevant, as all the information for the optics is contained in the 'kL' values. (**MAIN WINDOW.**) Click 'Save' and then 'Back' to return to the 'Tracking' window.

(6) Repeat (4) with −0.005 values for x and z. Select the entire field to be able to change the signs from the original values (this is but the first example of a bug in the GUI!—but be patient).

(7) Choose 'Trajectories' again in the 'Graphs' menu and then 'All.' The results show the wonders of 'alternating gradient' focusing: net focusing on both planes! One can now try increasing the separation of the quads to be more than their individual focal lengths. To do this, go back to the **MAIN WINDOW**, then select 'File', 'Edit'. Double-click on the field to be changed and press OK. Then go to 'Check Data' and select either option. It is now possible to repeat the tracking calculations.

A.5.4 Marylie

The use of Lie algebra methods for particle accelerator modeling and the MARYLIE code were conceived around 1973 by Prof. Alex Dragt, who was based at the University of Maryland for most of his career. Robert Ryne (Lawrence Berkeley National Laboratory (LBNL)), David Douglas (Jefferson Lab), and several others have contributed to code development over the years. From the MARYLIE 3.0 manual [8]:

'The program employs algorithms based on a Lie algebraic formulation of charged particle trajectory calculations and is able to compute transfer maps for and trace rays through single or multiple beamline elements. This is done for the full six-dimensional phase space without the use of numerical integration or traditional matrix methods. All nonlinearities, including chromatic effects, through third (octupole) order are included.'

We describe below the basic mechanics of running MARYLIE. Clearly, with our limited experience running the code, we cannot do justice here to the full capabilities of an advanced tool for particle accelerator design.

DOWNLOAD: the code is available as a module for the Particle Beam Optics Laboratory (*PBO Lab*) suite, a product of AccelSoft Inc. of Del Mar, CA [20]. Alternatively, the code is available at no charge to U.S. Government supported agencies or projects by contacting Prof. A Dragt (see table A.3).

INSTALLATION/RUNNING: if the FORTRAN code can be obtained, MARYLIE can be compiled in Ubuntu Linux or the Ubuntu app for Microsoft Windows 10 using the standard gfortran compiler. The command `gfortran -o ml3x.exe` followed by a small number of FORTRAN files listed in the file 'compileit' will yield an executable file. It is also possible to compile the code for MacOS.

EXAMPLE: chapter 2 of the MARYLIE manual [8] describes six examples: three simple systems and three rings. Chapters 6 and 10 discuss additional examples of beamline elements and special applications. The code distribution that we tried is from 2007; it contains folders labeled 'exh[chapter no.]_[example no.]'. Thus, the example illustrated in section 1.4 of the present book is found in folder 'exh02_3'. The 'master input file' **exh2_3.mif** should be renamed as fort.11 before running MARYLIE with ml3x.exe fort.11. In addition, an input data file fort.13 is needed; the original distribution had **exh2_3.ic**, which we have edited to represent the object shown in figure 1.5(a), including nine rays for every point in the object (combinations of initial slopes $(x'_0, y'_0) = \pm 2$ mrad, 0.0 mrad). The object is to be imaged by four quadrupole doublets by means of 20 TeV protons.

The structure of the input file is described in detail in the MARYLIE manual. We just point out here that the last two entries for the quadrupoles are set to +1.0 to indicate that the lenses have hard-edge leading and trailing fringe fields (see also the last paragraph in section B.4 of appendix B). The output file `fort.12` contains the input file as well as the 6×6 matrix for the map and the nonzero elements of the generating polynomial (section A.2). The output *data* is contained in the file `fort.14;` these are the numbers employed to plot figure 1.5(b).

To third order, a total of 209 monomials form the polynomials f_2, f_3, f_4. Note that monomials of up to *degree 4* contribute to *third-order* effects (chapter 3 of [8]). The nonzero coefficients of these monomials are listed at the end of `fort.12`; some take large values when fringe fields are included, indicating the presence of significant aberrations. For example, `f(140)=f(04 00 00)=-13424.25` denotes a large coefficient for the monomial P_x^4 (P_x is the conjugate momentum for the coordinate 'X'). When the quadrupole fringe-field parameters are set to zero, the same coefficient $f(140)$ is greatly reduced but remains nonzero because of the higher-order consequences of 'symplectification' [8].

A.6 Envelope codes

A.6.1 SPOT and MENV

The SPOT code solves the envelope equations (4.46) for either a unit FODO cell or a general matching section. In the first case, the periodic envelope solution is found for given beam parameters (generalized perveance and edge emittances) and fixed quadrupole strengths corresponding to the desired zero-current phase advances (section 7.1). These quadrupole strengths can be found using equations (3.16) and (3.17) for hard-edge gradient models. The other, perhaps more interesting, case of matching is the optimization of a nonperiodic envelope leading from the input plane at $s = s_i$ to the periodic envelope starting at $s = s_f$ (section 7.2). A major feature of SPOT is the use of a *reference trajectory*. If the average values of the beam excursions X and Y are known, the reference trajectory or trajectories can be set to be equal to those constant values, except at the beginning of the matching section (see figure 7.4). With a specified reference trajectory and an initial guess for the strengths of the matching lenses, SPOT minimizes a functional, which is essentially the squared distance between the envelope solution and the reference trajectory. This functional can be extended to include weighted beam target parameters, e.g., the beam transverse semi-axes and slopes at the terminal state. The optimization techniques in SPOT utilize tools from nonlinear programming developed for control problems in engineering and mathematics. Additional details can be found in [21, 22].

SPOT runs under Microsoft Windows 3.1 or Windows XP, or through emulation software such as *Wine* (Linux and Mac OSX). Dr. Hui Li, a former graduate student at University of Maryland, wrote a *Matlab* version of SPOT with a GUI and added functionality. The *Matlab* version is called MENV, which stands for '*Matlab* Envelope.' The optimization in MENV directly uses Matlab tools, which in some cases leads to solutions that converge more rapidly than in SPOT.

Typical workspaces for SPOT and MENV are shown in figure A.4.

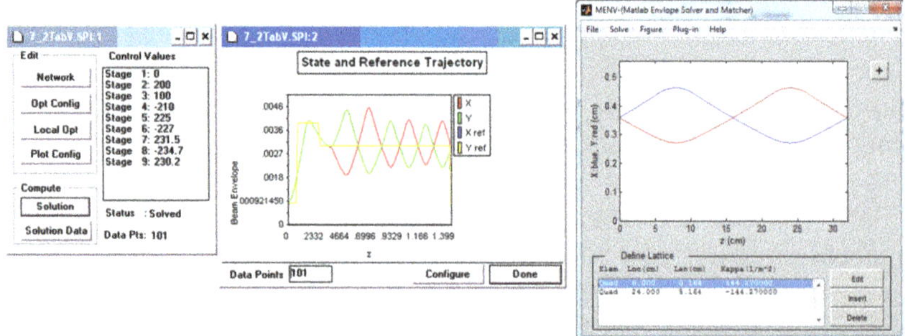

Figure A.4. Working windows in the envelope codes SPOT and MENV. The reference trajectory is shown in yellow for the nonperiodic matching problem in SPOT (middle plot). The MENV window (right) displays an example of periodic envelope matching. The examples are taken from UMER. See [21, 22].

Both SPOT and MENV can be obtained through this book's website.

A.6.2 TRACE 2-D and TRACE 3-D

The best description of the TRACE programs can be found in the abstract to the Trace 3-D documentation [23]: 'TRACE 3-D is an interactive beam-dynamics program that calculates the envelopes of a bunched beam, including linear space-charge forces, through a user-defined transport system. TRACE 3-D provides an immediate graphics display of the beam envelopes and phase-space ellipses and accommodates fourteen different types of fitting or beam matching options.' The documentation includes several appendices that include excellent concise discussions of basic topics such as the σ-matrix, Twiss parameters, mismatch factor, etc. Furthermore, a summary of TRACE 2-D commands is given in appendix K. TRACE 2-D is used for unbunched, i.e., continuous or coasting beams.

DOWNLOAD: the codes are freely available from Los Alamos National Laboratory after a simple registration process. See table A.3.

INSTALLATION: the .zip file is decompressed to reveal the executable file 'trace2d.exe' or 'trace3d.exe' which can be moved to the desktop in Windows. The examples and input files can remain in any folder, but the Windows system folder 'Program Files (x86)' is not recommended.

RUNNING / EXAMPLE: clicking on the TRACE icon produces a file-opening dialog. In our example, the input file is 'fodotest.t2d' (TRACE 2-D). A blank page is produced after opening the file, but simply typing 'g' (for 'graph') will render the beam horizontal and vertical phase-space ellipses at the input (left pane) and output (right pane) planes, together with the beamline and the transverse beam envelopes at the bottom. See figure A.5.

```
&DATA
ER= 0.511 Q= -1. W= 0.010 XI= 7.000
EMITI = 16 16
BEAMI = -0.914037 1.8 0.914037 1.8
BEAMF = 0.00000 1.8 0.00000 1.8
BEAMCI= 0.00000 0.00000 0.00000 0.00000
FREQ= 425.000 PQEXT= 2.50 ICHROM= 0 IBS= 0 XC= 0.0000
XM= 10.0 XPM= 50.0000 YM= 10.00
XMI= 5.0 XPMI= 50.0000 XMF= 5.0 XPMF= 50.0000
N1=1 N2=5 SMAX=0.5 PQSMAX=2.0 nel1=1 nel2=5 NP1=1 NP2=5
MT= 1 NC= 4 MP=0,000 0,000 0,000 0,000
MVC=0,000,0 0,000,0 0,000,0 0,000,0


CMT(001)='DRIFT' NT(001)= 1 A(1,001)= 54.18
CMT(002)='Quad' NT(002)= 3 A(1,002)= 0.57049E-01 51.64
CMT(003)='DRIFT' NT(003)= 1 A(1,003)= 108.36
CMT(004)='Quad' NT(004)= 3 A(1,004)= -0.57049E-01 51.64
CMT(005)='DRIFT' NT(005)= 1 A(1,005)= 54.18
```

```
WS(1) = 0.00000 0.00000 0.00000 0.00000 0.00000 0.00000 0.00000
WS(8) = 0.00000 0.00000 0.00000 0.00000 0.00000 0.00000 0.00000
SIGI(1,1) = 45.216 -186.84 0.00000E+00 0.00000E+00 0.00000E+00
SIGI(1,2) =-186.84 851.67 0.00000E+00 0.00000E+00 0.00000E+00
SIGI(1,3) =0.000E+00 0.000E+00 34.734 57.21 0.0000E+00
SIGI(1,4) =0.000E+00 0.000E+00 157.21 815.21 0.0000E+00
SIGI(1,5) =0.000E+00 0.000E+00 0.000E+00 0.000E+00 131.21
COMENT='FODO match for 7.0 mA, 10keV'
&END
```

The second line of 'fodotest.t2d' defines the beam: 'ER' is the rest mass in MeV, 'W' the kinetic energy in MeV, and 'XI' the beam current in mA (continuous beam). The next line contains the initial transverse edge emittances (EMIT) in mm-mrad. BEAMI (BEAMF) contains the initial (final) Courant–Snyder (Twiss) parameters α and β in the horizontal and vertical planes. XMI, XPMI, XMF, and XPMF define the scales (in mm) for the initial and final phase-space ellipses. XM, YM define the scales for the envelopes. There are five elements in the beamline, three drift spaces

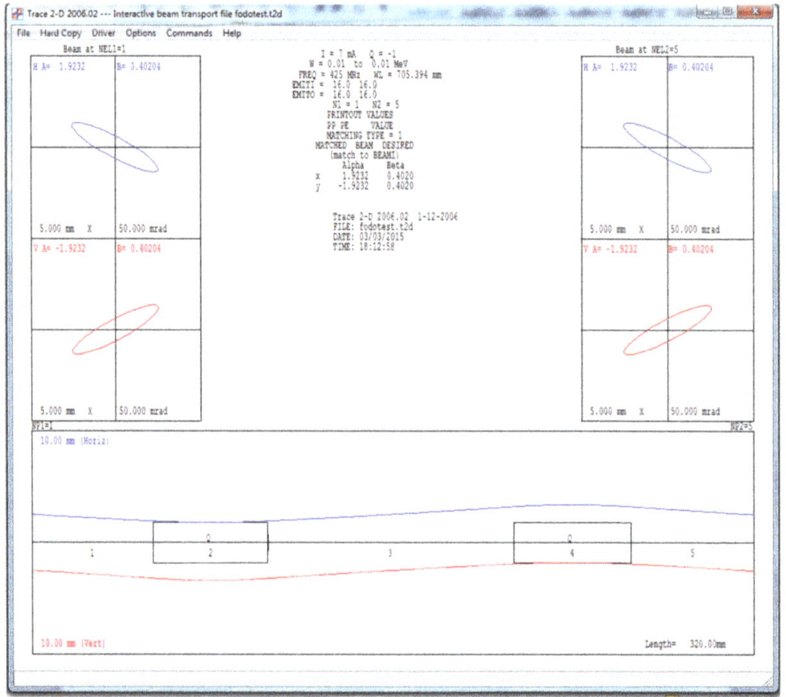

Figure A.5. TRACE 2D window after matching the beam envelopes in one cell of UMER for a 10 keV, 7 mA, 16 μm (x and y edge emittances) electron beam.

and two quadrupoles; the drifts are in mm, and the quadrupole parameters are the gradient in T/m and the hard-edge effective length (51.64 mm in our case).

The matching type is MT = 1 for ellipse parameters α_x, β_x, α_y, β_y. Envelope matching is carried out by typing 'm' and then 'g'. The phase advance and Courant–Snyder parameters are displayed by typing 'f'. A list of commands is provided in a dropdown menu. The input file can be modified and run again without saving or exiting the program. As an example, the current can be set to zero by typing 'i,' and then 'xi=0' in the new window; note that the beam is mismatched. After re-matching, new phase advances can be obtained; comparing these values with those obtained previously for full beam current illustrates the effect of linear transverse space charge (see chapter 4).

A.7 Particle-in-cell codes

Particle-in-cell (PIC) codes are widely used to simulate plasmas and beams; some incorporate interactions with intense lasers. PIC codes are discussed in detail in [24], which is a classic textbook for plasma simulations. Two basic features characterize PIC codes: (1) the use of a *spatial grid* for calculations of charge and current densities and electromagnetic fields; and (2) the use of *macroparticles*, which represent a large number of actual particles in the beam or plasma. The resolution of the grid is normally set to be smaller than one *Debye length* (section 4.4). A one-dimensional example of a PIC simulation is given in chapter 3 of [24]. An implementation in Mathcad is given in the file **13_PIC_code_ESwave.xmcd**, originally developed at the University of Colorado, Boulder; the file can be downloaded from this book's website at https://www.terpconnect.umd.edu/~sabern/PIBPPA-3rdEd/Appendix%20A/. Although the algorithms used to solve the equations of motion may themselves be symplectic (section A.1), the use of a spatial grid makes PIC codes inherently non-symplectic.

A.7.1 Warp

WARP is a hybrid plasma/accelerator PIC code created in the 1980s by Alex Friedman at the Lawrence Livermore National Laboratory; it includes contributions to code development made by David Grote (LLNL/LBNL) and many others over the years [25]. A good overview of WARP, including examples, can be found in the June 2016 USPAS presentations by J L Vay [26, 27]. The name 'Warp' refers to a special coordinate system used in bends. WARP is written in FORTRAN and is run using Python scripts that link to special compiled modules.

DOWNLOAD: the WARP website at LBNL is very comprehensive (see table A.3) and has clear instructions that can be used to download and compile the code.

INSTALLATION: We have compiled the code for Ubuntu Linux 16.04 using the standard gfortran compiler without any insurmountable problems. Installation of WARP is also possible on MacOS.

RUN/EXAMPLE: The June 2016 USPAS presentation 'W3. Examples WARP Simulations' by J L Vay [27] contains four examples with simulation exercises.

The Python scripts for all but the laser plasma acceleration example can be found in the 'examples' folder that is created by the WARP installation. The first example is a 93 kV singly-charged potassium-ion source consisting of a 5.5 cm radius hot plate, a Pierce focusing electrode (used to compensate for initial space-charge defocusing) and an anode extractor. The Python script **Pierce_diode.py** can be run in interactive or non-interactive mode, as described on the WARP website (table A.3). The default simulation employs an *r-z geometry* to calculate the time evolution of 150 macroparticles under space-charge limited conditions, i.e., Child–Langmuir emission (see chapter 2 in [28]). The simulation runs over 13 cm from the hot plate, or 5 cm beyond the anode aperture ('extractor'). The initial particle distribution is semi-Gaussian, and the field grid dimensions are $(nx, ny, nz) = (64, 64, 32)$. All realistic boundary conditions are implemented.

The time step is chosen to satisfy the *Courant limit*, or Courant–Friedrichs–Lewy (CFL) condition, which states that [29] 'for a given grid size Δx the time step Δt must be small enough that no signal can travel more than one grid cell in a single time step: $\Delta t \leqslant \Delta x/v$.' The CFL condition is sometimes expressed in terms of the plasma frequency:

$$\Delta t \leqslant \frac{2}{\omega_p}, \tag{A.10}$$

where ω_p is the plasma frequency (equation (4.12)). In the example discussed, the Python code includes the line `top.dt = 0.4*(dz/vzfinal)`, where `dz = 13cm/32 = 0.41cm` and `vzfinal = sqrt(2.*diode_voltage*jperev/beam.mass)`. In addition, it is interesting to note that just 150 macroparticles suffice for accurate simulation of the Pierce diode. This is possible because WARP calculates the fields, in this case, through a standard relaxation technique as used in the classical electron gun code EGUN [30], but particles are moved using the standard leapfrog method. The *Mathcad* file **Pierce-Diode WARP.xmcd** contains explicit calculations of the Child–Langmuir ion current, the Courant limit, the plasma frequency, and the 'weight of the simulation particles,' i.e., the number of real particles per macroparticle.

The results of current evolution as a function of axial distance are saved every 20 steps in a 'computer graphics metafile' (.cgm). The graphics/plots can be displayed using 'gist.' Several commands within 'gist' allow the conversion of multipage .cgm files into postscript format. Figure A.6 shows the output after 300 steps.

A.8 Mobile applications

Android: from the Google Play website—'TAPAs stands for Toolkit for Accelerator Physics on Androids. It allows "back-of-the-envelope" accelerator physics calculations, with more than 20 inter-linked types of calculations. These include storage ring scaling with energy and circumference, longitudinal dynamics in rings, free-electron lasers, electron linear accelerators, electron guns, bunch compression, iron-dominated multipole magnets, short pulse x-rays, undulator effects on orbit and optics, undulator

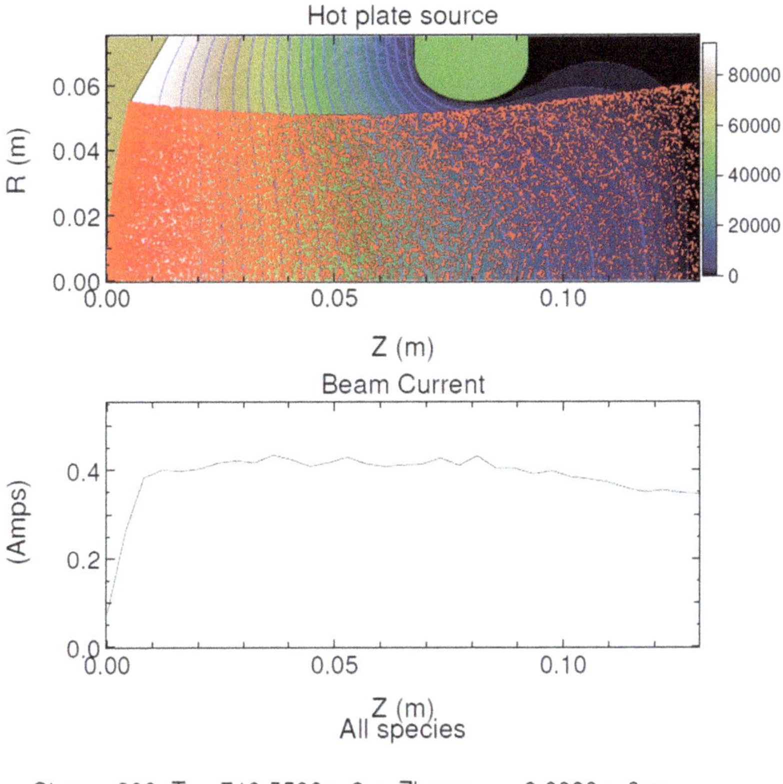

Figure A.6. Example of WARP output produced by simulating a 93 kV potassium-ion source. The 5.5 cm radius (curved) hot plate is on the left, and the anode extractor is represented by the green shape at 0.08 m. The red points are particles, while the color scale indicates electrical potential. See also [27].

radiation properties, undulator models, top-up parameters, electromagnetism, synchrotron radiation, particle passage through matter, and engineering calculations.' See also [31]. Figure A.7 shows a typical screenshot from TAPAS.

A.9 Cloud computing

RadiaSoft, LLC based in Boulder, CO has created SIREPO, a browser-based simulation 'gateway' that allows the user to run particle accelerator codes such as ELEGANT, MAD-X, SYNERGIA, and WARP without any installation. SIREPO provides a GUI similar to that of PBO Lab [20], which has multiple options for creating, editing, and sharing simulations. A video tutorial that describes how to run ELEGANT can be accessed at https://www.radiasoft.net/blog/what-is-sirepo-software-science-codes-overview/. The basic SIREPO is free, and commercial licenses are available for additional functionality and services.

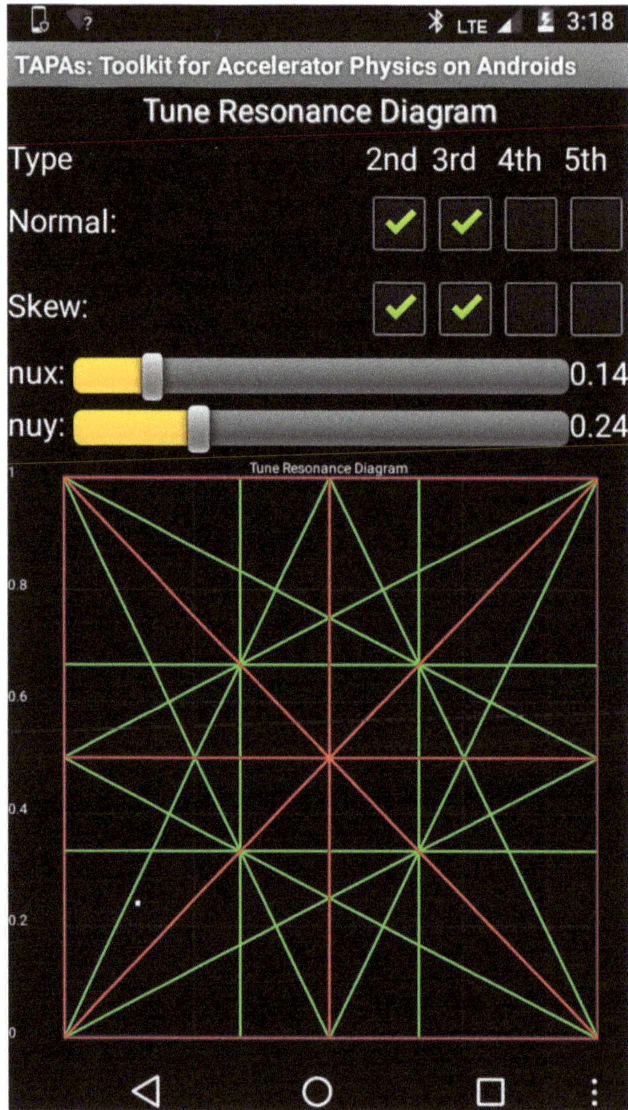

Figure A.7. Screen from Android app TAPAS.

References

[1] Goldstein H 1980 *Classical Mechanics* 2nd edn (Reading, MA: Addison-Wesley)

[2] Carey D C, Brown K L and Rothacker F 1995 Third-order TRANSPORT—a computer program for designing charged particle beam transport systems, SLAC-R-95–462, Fermilab-Pub-95/068, UC-414

[3] Dennis Donnelly and Edwin Rogers 2005 Symplectic integrators: an introduction *Am. J. Phys.* **73** 938

[4] Yoshida H 1999 Symplectic Integration Methods *Handbook of Accelerator Physics and Engineering* ed A W Chao and M Tigner 3rd printing (Singapore: World Scientific) 2.3.8 pp 101–3

[5] Étienne Forest 1998 *Beam Dynamics—A New Attitude and Framework* 1st ed. (Amsterdam: Hardwood Academic Publishers)

[6] Wolski A 2014 *Beam Dynamics In high Energy Particle Accelerators* (London: Imperial College Press)

[7] Dragt A J 2013 An overview of Lie methods for accelerator physics *Proc. of PAC 2013* (Pasadena, CA) 1134 http://accelconf.web.cern.ch/AccelConf/pac2013/papers/thap1.pdf

[8] Dragt A *et al* 2003 Marylie 3.0 users' manual *A Program for Charged Particle Beam Transport Based on Lie Algebraic Methods* (College Park, MD: University of Maryland)

[9] Dragt A J 1981 Lectures on nonlinear orbit dynamics *AIP Conf. Proc. No. 87*

[10] Dragt A J 2018 *Lie Methods for Nonlinear Dynamics with Applications to Accelerator Physics* (College Park, MD: University of Maryland) https://www.physics.umd.edu/dsat/dsatliemethods.html

[11] Cooper R K and Jones M E , The physics of codes *AIP Conf. Proc. No. 184* 1084

[12] Douglas D *et al* Simplified charged particle beam transport modeling using commonly available commercial software *Proc. PAC07 (New Mexico: Albuquerque)* 3651

[13] Grote H, Roy G, Schmidt F and Deniau L (ed) 2022 *The MAD-X Program (Methodical Accelerator Design), Version 5.08-01, User's Reference Manual* (Geneva Switzerland: European Laboratory for Particle Physics)

[14] Métral E 2013 *Training-week in Accelerator Physics* (Sweden: Lund)

[15] Murphy J 1993 Synchrotron Light Source Data Book, Version 3.0 October 1993, Brookhaven National Lab., BNL 42333 (revised) Informal Report.

[16] Borland M 2000 Technical Report No. LS-287, Argonne National Laboratory.

[17] Borland M 2021 User's manual for elegant, program version 2021.4 (Lemont, IL: Advanced Photon Source) November 19, 2021

[18] Bryant P J 2000 AGILE, A tool for interactive lattice design *Proc. EPAC (Vienna, Austria)* 1357

[19] Bryant P J and Johnsen J 1993 *The Principles of Circular Accelerators and Storage Rings* (Cambridge: Cambridge University Press)

[20] PBO Lab 1997 (Particle Beam Optics Laboratory) User Manual, AccelSoft, Inc, G.H. Gillespie Associates, Inc.

[21] Allen C K, Guharay S K and Reiser M 1996 Optimal transport of low energy particle beams *Proc. of the 1995 Particle Accelerator Conf., Dallas, Texas* (New York: IEEE) p 2324

[22] Bernal S, Li H, Kishek R A, Quinn B, Walter M, Reiser M and O'Shea P G 2006 RMS envelope matching of electron beams from 'zero' current to extreme space charge in a fixed lattice of short magnets *Phys. Rev. ST Accel. Beams* **9** 064202

[23] Crandall K R and Rusthoi D P 1997 TRACE 3D documentation, LA-UR-97-886 *Los Alamos National Laboratory Report* 3rd edn

[24] Birdsall C K and Langdon A B 2004 *Plasma Physics via Computer Simulation* (Boca Raton, FL: CRC Press)

[25] Friedman A, Cohen R H, Grote D P, Lund S M, Sharp W M, Vay J-L, Haber I and Kishek R A 2014 Computational methods in the warp code framework for kinetic simulations of particle beams and plasmas *IEEE Trans. Plasma Sci.* **42** 1321

[26] Vay J-L 2016 W2. introduction to the warp framework, U. S. Particle Accelerator School, Colorado State U., Ft. Collins, CO, 13–17 June 2016

[27] Vay J-L 2016 W3. Examples warp simulations, U. S. Particle Accelerator School, Colorado State U., Ft. Collins, CO, 13–17 June 2016.

[28] Reiser M 2008 *Theory and Design of Charged Particle Beams* 2nd edn (Weinheim: Wiley-VCH)

[29] Miloch W J 2014 *Plasma physics and numerical simulations* (Oslo: University of Oslo) https://www.uio.no/studier/emner/matnat/math/MEK4470/h14/plasma.pdf

[30] Hermannsfeldt W B 1979 Electron trajectory program *SLAC Report* 226

[31] Borland M Android application for accelerator physics and engineering calculations *Proc. of the 2013 Particle Accelerator Conf. (Pasadena, CA)* 1364

IOP Publishing

A Practical Introduction to Beam Physics and Particle Accelerators (Third Edition)

Santiago Bernal

Appendix B

Accelerator magnets

There is a vast literature on magnets for particle accelerators, electron microscopes, mass spectrometers, and other related devices (see, for example, the references in [1]). Many textbooks have either chapters or appendices that discuss the theory and design of accelerator magnets, e.g., [2–6]. The *Handbook of Accelerator Physics and Engineering* [7] also provides practical information in many specialized articles (see, for example, section 7.2.4 of that book. Finally, the manuals for particle accelerator codes such as MAD-X, ELEGANT, and others include useful, self-contained discussions and important sign conventions. Here, we present a brief account of magnet theory based on some of these references and on material from CERN and the US Particle Accelerator Schools. We concentrate on the basic theory of air-core magnets and refer the reader to the references for discussions of magnet design, calculations, and measurements including resistive and superconducting magnets.

In the first section, we introduce the multipole expansion of the transverse 2D B-field inside a magnet, the expansion in the midplane of combined-function magnets, and the K_n coefficients used in computer codes. In section B.2, we briefly discuss dipole and quadrupole magnets and their generation with $\cos(n\theta)$ current geometries. In section B.3, we describe sextupole and octupole magnets and their use, and in section B.4 we illustrate a technique that can be used to find the effective length of focusing magnets, which is especially useful when the magnets have significant fringe fields. The last section contains computer resources.

B.1 Multipole expansion

The transverse components of the B-field inside a magnet of circular cross section of radius r_0 (or the 'reference' radius inside the magnet) and far from its ends have the 2D multipole expansion representation

doi:10.1088/978-0-7503-4039-7ch11

$$B_y + iB_x = \sum_{n=1}^{\infty}(b_n + ia_n)\left(\frac{x + iy}{r_0}\right)^{n-1} =$$
$$\sum_{n=1}^{\infty}|C_n|\,\exp\left(in\phi_n\right)\left(\frac{r}{r_0}\right)^{n-1}\exp\left[i(n-1)\theta\right], \tag{B.1}$$

where $r = \sqrt{x^2 + y^2}$, $r < r_0$; b_n and a_n are the *normal* and *skew* $2n$-pole components, respectively; and $|C_n| = \sqrt{b_n^2 + a_n^2}$. In particular, $n = 1, 2, 3, \dots$ correspond to the *dipole, quadrupole, sextupole, ...* terms, respectively. Note that ϕ_n gives the orientation of the field with $\phi_n = 0$ for a normal multipole, and $\phi_n = \pi/2n$ for a skew multipole. We follow the 'European' convention in equation (B.1); in the 'US' convention, the summation starts at $n = 0$, so the power of $x + iy$ is 'n' instead of '$n-1$'. For example, b_1, a_1 in the 'European' notation correspond to b_0, a_0 in the 'US' convention. Thus, in general,

$$b_n(\text{Eur}) = b_{n-1}(\text{US}). \tag{B.2}$$

Of course, 'n' takes the same value on both sides in equation (B.2), starting at $n = 1$, as b_{-1} (US) would be meaningless.

From $B_y + iB_x = (B_\theta + iB_r)\exp(-i\theta)$, the expansion in equation (B.1) can be rewritten in cylindrical B-field components as

$$B_\theta + iB_r = \sum_{n=1}^{\infty}|C_n|\,\exp\left(in\phi_n\right)\left(\frac{r}{r_0}\right)^{n-1}\exp\left(in\theta\right), \tag{B.3}$$

or, explicitly in terms of b_n, a_n,

$$B_r(r, \theta) = \sum_{n=1}^{\infty}\left(\frac{r}{r_0}\right)^{n-1}[b_n\sin(n\theta) + a_n\cos(n\theta)], \quad r < r_0,$$
$$B_\theta(r, \theta) = \sum_{n=1}^{\infty}\left(\frac{r}{r_0}\right)^{n-1}[b_n\cos(n\theta) - a_n\sin(n\theta)], \quad r < r_0. \tag{B.4}$$

Figure B.1 illustrates the coordinate variables in the middle plane of the magnet bore.

The complex expansions in equations (B.1) and (B.3) are valid since the 2D B-field can be derived from Laplace's equation $\nabla^2\Phi = 0$ for a magnetic scalar potential Φ in the absence of current sources and the general solution for Φ written in the form of a complex analytic function. In section 2.3, we gave a real multipole expansion of Φ, from which the expansions of B_r, B_θ in equation (B.4) can be obtained from

$$B_r = -\frac{\partial\Phi}{\partial r}, \quad B_\theta = -\frac{1}{r}\frac{\partial\Phi}{\partial\theta}, \tag{B.5}$$

if we realize that $nd_n = b_n/r_0^{n-1}$, $nc_n = -a_n/r_0^{n-1}$, connecting the multipole coefficients in equations (2.12) and (B.1). Furthermore, the same expansions in

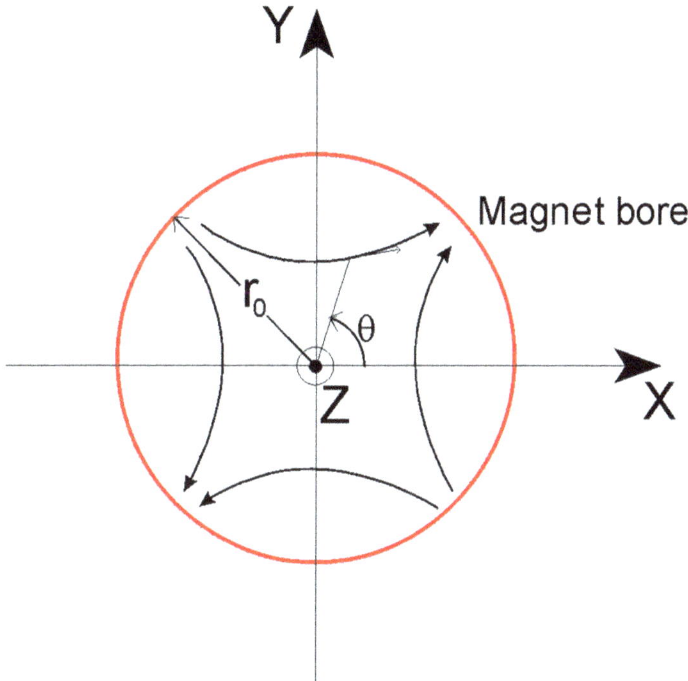

Figure B.1. Cylindrical coordinate system used for the magnet B-field description.

equations (B.1) and (B.4) are valid inside a magnet of *any* aspect ratio if one interprets B_x and B_y, or B_r and B_θ, as the *axially integrated* components of the B-field [8].

A well-designed $2n$-pole magnet has an n-multipole that is dominant by orders of magnitude relative to the 'allowed' and 'forbidden' higher and lower multipoles. Since a 180° rotation of a dipole is equivalent to inverting its polarity (see equation (B.3)), a dipole magnet can have multipoles given by $n = 2k + 1$ ($k = 0, 1, 2 \ldots$), i.e., *dipole* ($n = 1$), *sextupole* ($n = 3$), etc. Similarly, since a 90° rotation of a quadrupole is equivalent to inverting its polarity, a magnet with quadrupole symmetry has only multipoles of the form $n = 4k + 2$ ($k = 0, 1, 2, \ldots$), i.e., *quadrupole* ($n = 2$), *duodecapole* ($n = 6$), *10-pole* ($n = 10$), etc. The same argument can be applied to nonlinear magnets such as the sextupole and octupole magnets discussed later.

Using equation (B.1), we can write the first few terms of the expansions of B_y and B_x as follows:

$$B_y(x, y) = b_1 + b_2\frac{x}{r_0} + b_3\frac{1}{r_0^2}(x^2 - y^2) + b_4\frac{1}{r_0^3}(x^3 - 3xy^2) - a_2\frac{y}{r_0} - a_3\frac{2xy}{r_0^2}$$

$$+ a_4\frac{1}{r_0^3}(y^3 - 3x^2y) + \ldots, \tag{B.6}$$

$$B_x(x, y) = a_1 + a_2\frac{x}{r_0} + a_3\frac{1}{r_0^2}(x^2 - y^2) + a_4\frac{1}{r_0^3}(x^3 - 3xy^2) + b_2\frac{y}{r_0} + b_3\frac{2xy}{r_0^2}$$
$$- b_4\frac{1}{r_0^3}(y^3 - 3x^2y) + \ldots. \tag{B.7}$$

On the midplane of a magnet, $y = 0$, and B_y depends only on x and the b_n coefficients. If, in addition, $a_n = 0$ for all n, we have a *normal* magnet for which $B_x = 0$, and we can rewrite the expansion in equation (B.6) as

$$B_y(x, 0) = (B_0\rho)\sum_{n=0}^{\infty} K_n\left(\frac{x}{r_0}\right)^n, \quad K_n = \frac{1}{n!(B_0\rho)}\left(\frac{\partial^n B_y}{\partial x^n}\right)_{x=0=y}. \tag{B.8}$$

In equation (B.8), $(B_0\rho)$ is the magnetic rigidity. Comparing equations (B.6) and (B.8), we can write $b_{n+1} = (B_0\rho)K_n$, $n = 0, 1, 2, \ldots$. Note that we have changed the start value of the index n; we have done this to simplify the notation and to match the labeling of the K_n coefficients in computer codes such as MAD-X, TRANSPORT, ELEGANT, and many others (see appendix A for descriptions and references). An expansion similar to equation (B.8) applies to a *skew* magnet, for which $b_n = 0$ for all n in equation (B.7), so the field is purely horizontal on the midplane $y = 0$.

B.2 Linear magnets

Magnets for which the expansions in equations (B.6) or (B.8) contain more than one n-multipole are called *combined-function* magnets. Typically, dipole and quadrupole functions are combined, i.e., b_1, $b_2 \neq 0$ or K_0, $K_1 \neq 0$. We have

$$K_0 = \frac{B_y}{(B_0\rho)}, \quad K_1 = \frac{1}{(B_0\rho)}\frac{\partial B_y}{\partial x}. \tag{B.9}$$

As illustrated in figures B.2(a) and (b), displaced, overlapping (and infinite in the z-direction) solid elliptical cylinders carrying uniform current density generate pure multipole fields. For a dipole magnet, this result can be derived from the B-field of a solid conductor of elliptical cross section and infinite length; the superposition of the two conductors cancels the current at the intersection and leads to a pure normal dipole field [8]

$$B_y = \frac{\mu_0 J b x_0}{(a + b)}, \quad B_x = 0, \tag{B.10}$$

where J is the (constant) current density (in A m^{-2}), a and b are the horizontal and vertical semi-axes, and x_0 is the distance between the centers of the elliptical cylinders. For the quadrupole case, the complex B-field is [8]

$$B_y + iB_x = \frac{\mu_0 J(a - b)}{(a + b)}(x + iy). \tag{B.11}$$

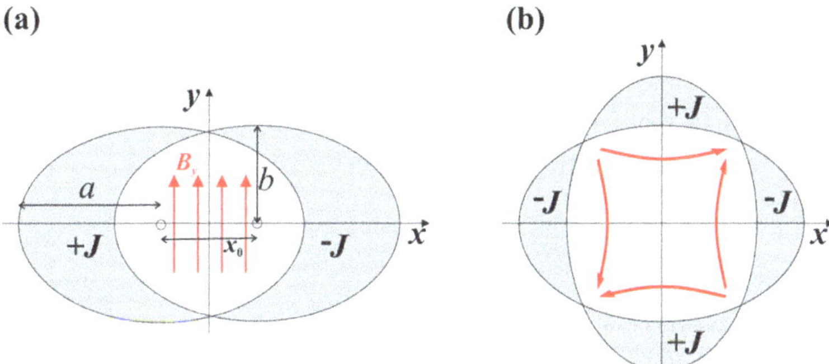

Figure B.2. Overlapping solid elliptical cylinders generate pure (a) dipole and (b) quadrupole fields. J represents a uniform current density in the $+z$-direction.

More generally, a current distribution of the form

$$I(\phi) = I_0 \cos(m\phi) \tag{B.12}$$

on a cylindrical surface generates a pure $2m$-pole B-field. We can make use of the complex B-field to prove this [8]. From Ampere's law, the B-field of an infinite filament carrying a current I along the $+z$-axis is $\boldsymbol{B} = (\mu_0 I/2\pi r)\hat{\boldsymbol{\theta}}$, where r is the radial distance from the filament, and $\hat{\boldsymbol{\theta}}$ is the unit vector in the azimuthal direction. If the filament is located at a point represented by the *complex* number $\mathbf{a} = a\exp(i\phi)$, the complex field at $z = x + iy$ is $B_y + iB_x = [\mu_0 I/2\pi(z - a)]$. A multipole expansion of this complex field for $r < a$, i.e., in the 'inside' region of the conductor is [8]

$$B_y + iB_x = -\left(\frac{\mu_0 I}{2\pi a}\right)\sum_{n=1}^{\infty} \exp(-in\phi)\left(\frac{r_0}{a}\right)^{n-1}\left(\frac{x + iy}{r_0}\right)^{n-1}. \tag{B.13}$$

A thin cylindrical shell of radius a with a current distribution $I(\phi)$ has a B-field inside it equal to

$$B_y + iB_x = -\left(\frac{\mu_0}{2\pi a}\right)\int_0^{2\pi} \sum_{n=1}^{\infty}\left(\frac{x + iy}{a}\right)^{n-1}[\cos(n\phi) - i\sin(n\phi)]I(\phi)d\phi. \tag{B.14}$$

Clearly, from the orthogonality of the sine and cosine functions, a current distribution of the form used in equation (B.12) yields a pure *normal* multipole of order $n = m$ with a complex field inside the shell given by

$$B_y + iB_x = -\left(\frac{\mu_0 I_0}{2a}\right)\left(\frac{x + iy}{a}\right)^{m-1}. \tag{B.15}$$

Despite their simplicity, neither the $\cos(n\phi)$ distribution nor the intersecting elliptical cylinders can be realized with sufficient accuracy in practice but are important first steps in magnet designs [8].

B.3 Lowest-order nonlinear magnets: sextupole and octupole magnets

The transverse Cartesian components of the B-field inside an ideal *normal* sextupole magnet are, according to equations (B.6) and (B.7),

$$B_y = \frac{b_3}{r_0^2}(x^2 - y^2), \quad B_x = \frac{2b_3}{r_0^2}xy. \tag{B.16}$$

The *normalized* sextupole strength or coefficient (in m^{-3}) is, according to equation (B.8),

$$K_2 = \frac{1}{2(B_0\rho)}\frac{\partial^2 B_y}{\partial x^2} = \frac{1}{(B_0\rho)}\frac{b_3}{r_0^2}. \tag{B.17}$$

Similarly, for a *normal* octupole magnet, we have

$$B_y = \frac{b_4}{r_0^3}(x^3 - 3xy^2), \quad B_x = \frac{b_4}{r_0^3}(y^3 - 3x^2y) \quad \text{and} \tag{B.18}$$

$$K_3 = \frac{1}{6(B_0\rho)}\frac{\partial^3 B_y}{\partial x^3} = \frac{1}{(B_0\rho)}\frac{b_4}{r_0^3}, \quad \text{in m}^{-4}. \tag{B.19}$$

Figure B.3(a) depicts the field of a normal sextupole magnet. Sextupole magnets are employed for chromaticity correction in storage rings; figure B.3(b) illustrates the basic idea. As we saw in section 3.6, natural chromaticity is of the order of the betatron tune and negative (see equation (3.48)). According to equations (B.16) and (B.17), the sextupole has gradients $\partial B_y/\partial x = \partial B_x/\partial y = (2b_3/r_0^2)x = 2(B_0\rho)K_2 x$. Thus, a sextupole placed next to a quadrupole in which the dispersion D_0 (section 3.5) is nonzero adds to the action of the quadrupole: particles with a fractional momentum deviation $\delta = \Delta p/p$ at a displacement $x = D\delta$ undergo the additional focusing (or defocusing if $\delta < 0$, as in figure B.3(b)) given by $2(B_0\rho)K_2 D\delta$. Integrating

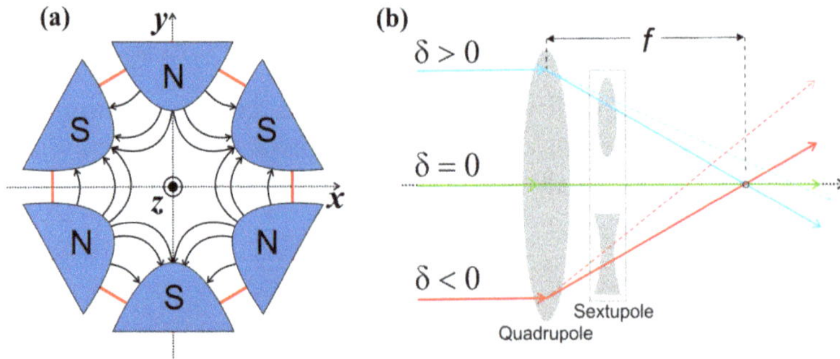

Figure B.3. (a) Schematics of a normal sextupole magnet. (b) Action of a quadruple–sextupole in a region of nonzero dispersion. $\delta = \Delta p/p$ is the fractional momentum deviation. Broken-line trajectories correspond to quadrupole focusing alone. See the text and [9].

over one turn and including the contributions of all sextupoles and quadrupoles, the total horizontal chromaticity is

$$\xi_{\text{Total }X} = \frac{\Delta\nu_X}{\delta} = -\frac{1}{4\pi}\oint[2K_2(s)D(s) + \kappa_X(s)]\beta_X(s)ds. \tag{B.20}$$

Properly placing and powering sextupoles around the accelerator can yield zero net chromaticity, although it is common to use small positive values.

Octupole magnets, on the other hand, introduce amplitude-dependent tunes, as in the quasi-integrable (QI) lattices in IOTA and UMER, which may help beam stability via Landau damping but at the cost of reduced dynamic aperture (chapter 9). Naturally, octupoles are also routinely used in large machines such as the LHC and are being considered for the Future Circular Collider (FCC). Calculations and simulations of beam stability for the FCC are discussed in [10].

B.4 Effective hard-edge model of fringe fields in focusing magnets

The standard *hard-edge model* of the focusing function $\kappa(s)$ was illustrated in figure 1.2. It has a length equal to the effective length as defined in equation (1.7) and a maximum value that coincides with the maximum value of $\kappa(s)$. However, the hard-edge model constructed in this way *will not* give correct answers in general. For this reason, other models have been devised to account for the smooth wings of an actual $\kappa(s)$ profile. One example is the linearization of the end fields. However, for basic design calculations it is still more convenient to implement hard-edge models in computer codes to track particle trajectories or trace beam envelopes.

The correct approach to constructing a hard-edge model consists in dividing the focusing function profile into a number of thin slices representing individual lenses and then multiplying out all the individual matrices to obtain a transfer matrix whose elements can be related to the effective length and strength of the magnet [5, 11]. This method of 'slicing' the focusing profile is quite general and is especially suitable for cases in which the profile has extended wings.

Here, we illustrate the slicing method using the focusing function for a solenoid. The function is proportional to the square of the on-axis z-component B_z of the solenoid B-field—see equation (2.11):

$$B_Z(s) = B_0 \exp\left(-\frac{s^2}{d^2}\right)\left[\text{sech}\left(\frac{s}{b}\right) + c_0 \sin h^2\left(\frac{s}{b}\right)\right], \tag{B.21}$$

where B_0 is the peak field, i.e., the axial field at the midplane of the solenoid ($z = 0$), and d, b, and c_0 are constants. Equation (B.21) can be used to very accurately fit actual magnetic field data for a solenoid used to focus low-energy electrons [12]. Figure B.4 illustrates the $\kappa(s)$ profile.

The profile in figure B.4 is divided into $N = 203$ slices with a width $\delta = 0.001$ m, thus covering $L = 0.203$ m well into the wings where $\kappa \ll 1\%\kappa_0$. Each slice is represented by the following matrix, (see equation (3.1) and also equations (1.1)–(1.3)):

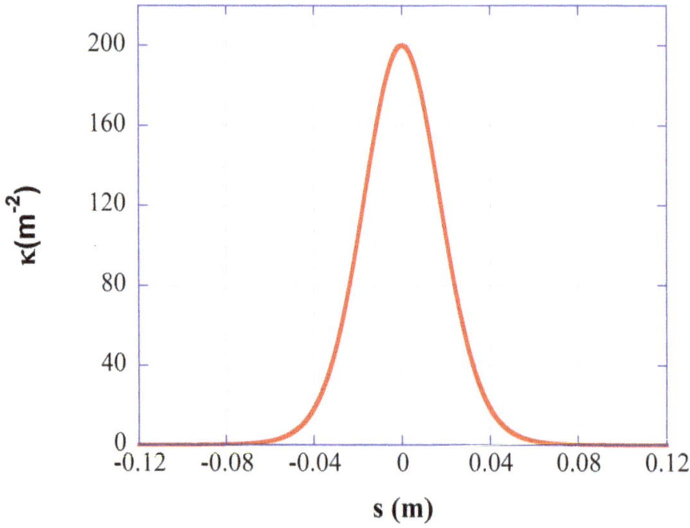

Figure B.4. Analytical axial profile of the focusing function for a short solenoid. The profile is proportional to the square of $B_z(s)$ in equation (B.21) with particular values of the fit constants and a peak value of $\kappa_0 = 200$ m^{-2}. See also the Mathcad document.

$$R_{\text{slice}} = \begin{bmatrix} \cos\left(\sqrt{|\kappa(s)|}\,\delta\right) & \dfrac{1}{\sqrt{\kappa(s)}} \sin\left(\sqrt{|\kappa(s)|}\,\delta\right) \\ -\sqrt{\kappa(s)}\,\sin\left(\sqrt{|\kappa(s)|}\,\delta\right) & \cos\left(\sqrt{|\kappa(s)|}\,\delta\right) \end{bmatrix}. \tag{B.22}$$

Multiplication of all N matrices leads to:

$$R = R_1 R_2 ... R_N = \begin{bmatrix} R_{11} & R_{12} \\ R_{21} & R_{22} \end{bmatrix}. \tag{B.23}$$

The *effective length* of the solenoid is then found using [5]:

$$l_{\text{eff}} = -\frac{\theta \sin\theta}{R_{21}}, \tag{B.24}$$

where θ is the root (nearest to zero) of the transcendental equation

$$\cos\theta + \frac{1}{2}\theta \sin\theta - R_{11} + \frac{1}{2}LR_{21} = 0. \tag{B.25}$$

'L' is the extent of the slicing. Furthermore, the *effective strength* of the solenoid is:

$$\kappa_{\text{eff}} = \frac{\theta^2}{l_{\text{eff}}^2}. \tag{B.26}$$

The method just described yields hard-edge models whose hardtop values generally *do not* coincide with the peak value of the smooth profile. This is illustrated in figure B.5. Furthermore, there is a slight dependence of the effective length and effective hardtop strength on the peak strength of the smooth profile $\kappa(s)$. Explicitly, we obtain $l_{\mathrm{eff}} = 0.0651$ m and $\kappa_{\mathrm{eff}} = 138.62$ m^{-2} for the profile illustrated in figure B.4, for which $\kappa_0 = 200$ m^{-2}. For completeness, the hard-edge magnet matrix must be multiplied on both sides by drift matrices whose extents equal [11]:

$$\lambda = \frac{1}{2}(L - l_{\mathrm{eff}}). \tag{B.27}$$

In the numerical example under discussion, we find $\lambda = 0.0689$ m, which is slightly larger than the effective length itself. Additional details can be found in the references as well as in the Mathcad documents mentioned in the last section.

A similar 'slicing' procedure is applicable to *quadrupole* lenses, but two slightly different effective lengths are obtained: one for the focusing plane (l_f) and another for the defocusing plane (l_d). Likewise, two effective quadrupole strengths result from the analysis. Thus, instead of equation (B.24), we have:

$$l_f = -\frac{\theta_f \sin \theta_f}{F_{21}}, \qquad l_d = -\frac{\theta_d \sin h\theta_d}{D_{21}}, \tag{B.28}$$

where θ_f, θ_d are solutions of the transcendental equations

$$\cos \theta_f + \frac{1}{2}\theta_f \sin \theta_f - F_{11} + \frac{1}{2}LF_{21} = 0,$$
$$\cos h\theta_d - \frac{1}{2}\theta_d \sin h\theta_d - D_{11} + \frac{1}{2}LD_{21} = 0. \tag{B.29}$$

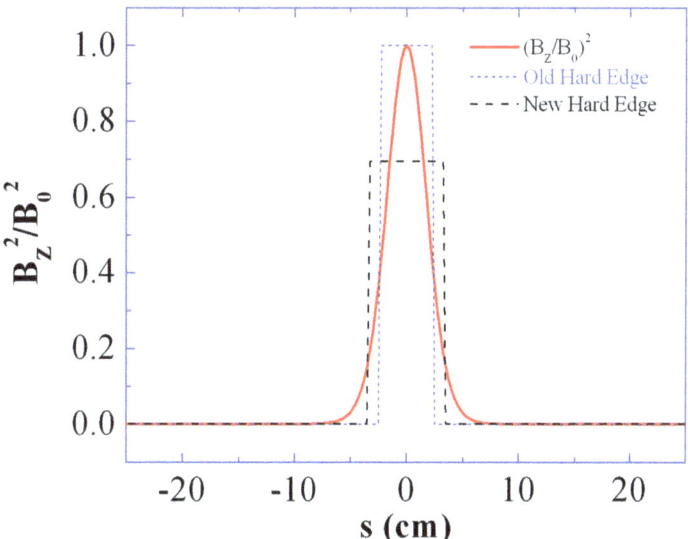

Figure B.5. Hard-edge models for the normalized focusing function of a short solenoid. The dotted blue line follows the prescription of equation (1.7), while the dashed black line results from the slicing method described in the text. From [12].

The corresponding effective focusing constants are related to θ_f, θ_d through $\theta = |\kappa|^{1/2} l$. The effective lengths l_f, l_d vary linearly with the quadrupole peak gradient in such a way that the *average effective length* $l_{\text{ave}} = (l_f + l_d)/2$ is nearly independent of the quadrupole gradient. An average effective quadrupole focusing constant is also defined. An example for short magnetic quadrupoles is illustrated in a Mathcad document available in this book's website.

The hard-edge model just described represents the correct *linear* (first-order) representation of a smooth focusing function; however, it is restricted to applications in which there is no significant field overlap at the fringe regions, as may occur when neighboring magnets are placed too close to each other. Field overlap would occur, for example, if the solenoid in the numerical example discussed above were part of a lattice with a half-period smaller than about $2\lambda + l_{\text{eff}}$, with λ as defined in equation (B.27).

A different, more general approach to 'hard-edge' models is provided by the map code MARYLIE (see section 1.4 and appendix A). Essentially, if ε is a parameter that characterizes the *fringe-field* region of a magnet, and $f_\varepsilon(z)$ represents the on-axis field or gradient, a 'hard-edge' map in the language of Lie maps is the result of taking the limit of the magnet map M_ε as $\varepsilon \to 0$. For quadrupole magnets, the resulting 'hard-edge' map M_0 retains the nonlinearities up to third order and is independent of the analytical model of the fringe-field region; for solenoid and dipole magnets, however, there is no hard-edge map beyond the second order [13].

B.5 Computer resources

The Hao Research Group of Accelerator Physics at Michigan State University provides useful material on magnets, including Python scripts, at https://people.nscl.msu.edu/~haoy/teaching/AP_course_materials/magnets/magnets.html.

Many computer codes are available for magnet design calculations, but the best are either expensive commercial products (e.g., OPERA 3D, MERMAID) or are limited to 2D or axially symmetric geometries (e.g., POISSON SUPERFISH, FEMM). There is also the free Python application Magpylib for calculations based on analytical formulas of permanent and air-core conductor magnets for different 3D geometries. At the UMER facility, we have used the commercial legacy code MAG-PC and our own Windows version MAG-Li to perform the design calculations for our air-core printed-circuit magnets [14, 15].

The Mathcad files **SOLENOID-SmoothAndHE.xmcd**, and **QUAD-SmoothAndHE.xmcd** illustrate the method described in section B.4 for calculating the effective lengths and strengths of solenoids and quadrupoles with fringe fields. Using these files, figures B.4 and B.5 for the solenoid case can be reproduced.

References

[1] Reiser M 2008 *Theory and Design of Charged Particle Beams* 2nd edn (Weinheim: Wiley-VCH)

[2] Bryant P J and Johnsen K 1993 *The Principles of Circular Accelerators and Storage Rings* (Cambridge: Cambridge University Press)

[3] Stephen Peggs and Todd Satogata 2017 *Introduction to Accelerator Dynamics* (Cambridge: Cambridge University Press)

[4] Rosenzweig J 2003 *Fundamentals of Beam Physics* (Oxford: Oxford University Press)

[5] Wiedemann H 2007 *Particle Accelerator Physics* 3rd edn (Berlin: Springer)

[6] Wolski A 2014 *Beam Dynamics in High Energy Particle Accelerators* (London: Imperial College)

[7] Wu Chao A, Mess K H, Tigner M and Zimmermann F 2012 *Handbook of Accelerator Physics and Engineering* 2nd edn (Singapore: World Scientific Publishing Company)

[8] Jain A K 1997 Basic theory of magnets (Anacapri, Italy) 1–26 CERN-98-05 Proc. CERN Accelerator School on Measurement and Alignment of Accelerator and Detector Magnets

[9] Wille K 2000 *The Physics of Particle Accelerators, an introduction* (Oxford: Oxford University)

[10] Kornilov V and Boine-Frankenheim O 2018 Landau damping and tune-spread requirements for transverse beam stability *Ninth Int. Particle Accelerator Conf. IPAC2018 (Vancouver, BC, Canada)* https://accelconf.web.cern.ch/ipac2018/papers/thpaf079.pdf

[11] Steffen K G 1965 *High Energy Beam Optics* (New York: Interscience)

[12] Bernal S *et al* 2006 'RMS envelope matching of electron beams from 'zero' current to extreme space charge in a fixed lattice of short magnets *Phys. Rev. ST Accel. Beams* **9** 064202

[13] Venturini M and Dragt A 2013 3D multipole expansion, calculation of transfer maps from field data, fringe fields *Handbook of Accelerator Physics and Engineering* ed A W Chao, K H Mess, M Tigner and F Zimmerman 2nd edn (Singapore: World Scientific) 2.2.2 pp 75–6

[14] Zhang W W, Bernal S, Li H, Godlove T, Kishek R A, O'Shea P G, Reiser M and Yun V 2000 Design and field measurements of printed-circuit quadrupoles and dipoles *Phys. Rev. ST Accel. Beams* **3** 122401

[15] Baumgartner H, Bernal S, Haber I, Koeth T, Matthew D, Ruisard K, Teperman M and Beaudoin B 2016 Quantification of octupole magnets at the University of Maryland Electron Ring *Proc. of NAPAC2016 (Chicago, IL)* 503–6 https://accelconf.web.cern.ch/napac2016/papers/tupob11.pdf